Science Strategies to Increase STUDENT LEARNING and MOTIVATION in Biology and Life Science Grades 7 through 12

DAVID BUTLER

CD ARTIST: ELLEN CORNWELL

Copyright © 2021 David Butler
All rights reserved
First Edition

PAGE PUBLISHING, INC.
Conneaut Lake, PA

First originally published by Page Publishing 2021

ISBN 978-1-6624-2657-5 (pbk)
ISBN 978-1-6624-2658-2 (digital)

Printed in the United States of America

To My Wife, Carol, for Her Support and Encouragement
and
To All Dedicated Teachers—for Without You, There Are No Other Professions

Contents

Notes 7

**Key Topics to Unlock to Increase Student Learning
and Motivation in Biology / Life Science** 9

Helping All Students Be More Successful in the Biology / Life Science Classroom 10
    Student Ownership 10
    Showcasing Work 19
    Providing Meaning 20
    Keeping It Professional 36
    Having an Open-Door Policy 49
    Checking for Understanding 55
    Having an Appropriate Atmosphere 65
    Having a Lesson Plan Outline 66
    Achievable Goals 68
    Varying Instruction 78

Increasing Student Motivation in the Biology / Life Science Classroom 82
    Enthusiasm 82
    Showmanship 83
    Feedback for Progress 83
    Reasons to Be Successful 85
    Topic Immersion/Focus 86
    Student Participation 87
    Encouragement and Support 88
    Highlighting Victories 89
    Finding Their Interests 90
    Going Outside the Classroom 93

**Incorporating Next-Generation Science Standards to Enhance Biology / Life Science
Instruction** 94
    "How Can There Be So Many Similarities among Organisms yet So Many
    Different Plants, Animals, and Microbes?"...NGSS 94
    Addendum 100
    "How Do Organisms Obtain and Use Energy They Need to Live and Grow? How
    Do Organisms Interact with the Living and Nonliving Environment to Obtain
    Matter and Energy?"...NGSS 107

    "How Do the Structures of Organisms Enable Life's Functions?"…NGSS 112
    "How Are the Characteristics from One Generation Related to the Previous
    Generation?"…NGSS 117

Differentiating Instruction More Efficiently in the Biology / Life Science Classroom 125
    Major Learning Style Strategies 125
    Continual Assessment Strategies 127
    Flexible Grouping Strategies 127
    Element Strategies to Help Determine Learning Preferences, Interests, and Readiness 128
    Prep Time / Homework / Extra Credit Strategies 128

Making the Best Use of Technology to Enhance Biology / Life Science Instruction 129
    Hardware 129
    "Secure" Software/Sites/Apps 129
    Teacher Created Online Tools 132

Improving Understanding and Retention of Biology / Life Science Concepts 136
    Music and Dance 136
    Illustrations 136
    Self-Disclosure 137
    Puns and Humor 137
    Mimicry 137
    Mnemonics 138
    Familiarities/Analogies 138
    Writing Notes / Organization 139
    Study Skills 139
    Review Sessions 140

Promoting Greater Student Success by Managing the Biology / Life Science Classroom 146
    Information/Policies 146
    Avoid Having to Discipline 149
    Trust and Respect 150

Using Demos and Unusual Materials to Increase Understanding of Biology / Life
    Science Concepts 151

Philosophy of Failing 155
References 157
Notes 159

Notes

It may be assumed that, as a rule, the teacher understands the subject with which (they) are entrusted, and has mastered its content, but not that (they) know how to impart (their) information in an interesting manner. This is almost always the source of the trouble. If the teacher generates an atmosphere of boredom, the progress is stunted in the suffocating surroundings. To know how to teach is to be able to make the subject of instruction interesting, to present it, even if it happens to be abstract, so that the soul of the pupil resonates in sympathy with that of (their) instructor, and so that the curiosity of the pupil is never allowed to wane.

Albert Einstein explaining his concern about education in an interview in 1921

Key Topics to Unlock to Increase Student Learning and Motivation in Biology / Life Science

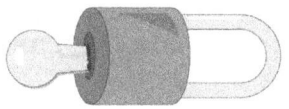
Lock-Key Complex

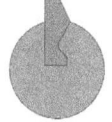
Enzyme-Substrate Complex

Helping All Students Be More Successful in Learning Biology / Life Science Concepts

Increasing Student Motivation for Learning Biology / Life Science Content

Incorporating Next-Generation Science Standards to Enhance Biology / Life Science Instruction

Differentiating Instruction More Efficiently in the Biology / Life Science Classroom

Making the Best Use of Technology to Enhance Biology / Life Science Instruction

Improving Understanding and Retention of Biology / Life Science Concepts

Promoting Greater Student Success by Managing Your Biology / Life Science Classroom

Using Demos and Unusual Materials to Increase Understanding of Biology / Life Science Concepts

Key (substrate)

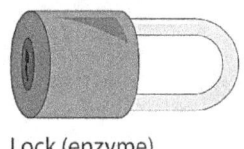
Lock (enzyme)

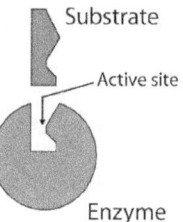
Substrate — Active site — Enzyme

Helping All Students Be More Successful in the Biology / Life Science Classroom

Helping Students Be More Successful with Student Ownership

- Pride of accomplishment
- Willingness to present
- Self-evaluation
- Sense of responsibility
- Empowers confidence
- Fosters metacognition (metaphorically, "driving one's own brain")

Strategy Examples of Student Ownership

- Chia Animals
- Decomposer: Fungus Lab
- Plant and Animal Diversity Teaching Assignment
- Human DNA Extraction Lab
- Cancer: Nine-Page-Spread Project
- Insect Recipe Lab
- BioBots: An Anatomical Collage of the Human Body

Introduction to botany "hook" mini-lab where students created their own chia animals

Addendum

Plant and Animal Diversity Teaching Assignment

Introduction

Biodiversity is the variation of life-forms within a given ecosystem, biome, or the entire earth. Biodiversity is often used as a measure of the health of biological systems. The biodiversity found on earth today consists of many millions of distinct biological species, which is the product of nearly 3.5 billion years of evolution. Two major kingdoms that contribute to the diversity of life on earth are plants and animals.

Plants have six fundamental characteristics: photosynthesis as the almost-exclusive mode of nutrition, essentially unlimited growth at meristems, cells that contain cellulose in their walls and are therefore somewhat rigid, the absence of organs of movement, the absence of sensory and nervous systems, and life histories that show alternation of generations. Many plants, for example, are not green and thus do not produce their own food by photosynthesis, being instead parasitic on other living plants. Others obtain their food from dead organic matter.

Animals have developed muscles, making them capable of spontaneous movement; more elaborate sensory and nervous systems; and greater levels of general complexity. Unlike plants, animals cannot manufacture their own food and thus have adapted for securing and digesting food. In animals, the cell wall is either absent or composed of material different from that of the plant cell wall. Animals account for about three-quarters of living species. Some one-celled organisms display both plant and animal characteristics.

Information/Requirements

Research plant and animal diversity from one of five preassigned (TBA) major biomes including aquatic, desert, forest, grassland, and tundra. One particular informative website about biomes would be https://ucmp.berkeley.edu/exhibits/biomes/index.php. How you wish to deliberate the research is up to you and your partner(s). Grading will be separate, however.

- Use presentation software to showcase your lesson. Suggestions include PowerPoint, Emaze, SlideRocket, Prezi, Google Presentation, 280 Slides, or Powtoon, just to name a few.
- Unless you purchased a membership, just use the free version of the program. It is also highly encouraged that you view any tutorials associated with the presentation program.

The presentation needs to be professional (correct grammar, pleasing font size/color, appropriate backgrounds/graphics/videos, not wordy, and proper beginning and endings). Be prepared to send or share your finished product via e-mail, shared folder, flash drive, network drive, etc.

- Teach (not just present) the topic by focusing on the following key requirements:
    - Know and understand your topic.
    - Develop a "hook" to begin.
    - Use a suitable manipulative(s).
    - Show enthusiasm for the topic.
    - Discuss rather than just reading.
    - Engage and mingle among the class.
    - Ask questions and check for understanding.
    - Address styles of learning (visual, auditory, kinesthetic).

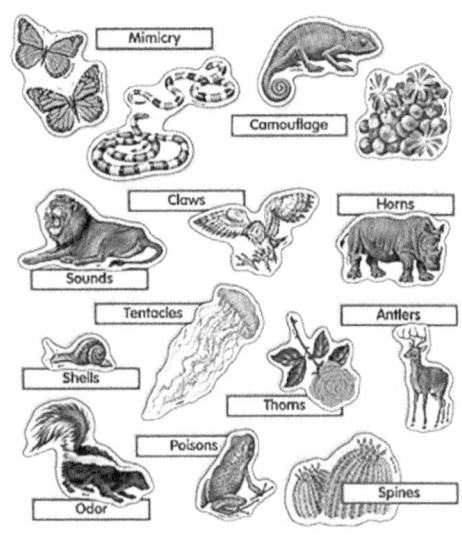

The entire group lesson is not to go over fifteen to twenty minutes from beginning to end. Be sure to review the *rubric* prior to your research and teaching.

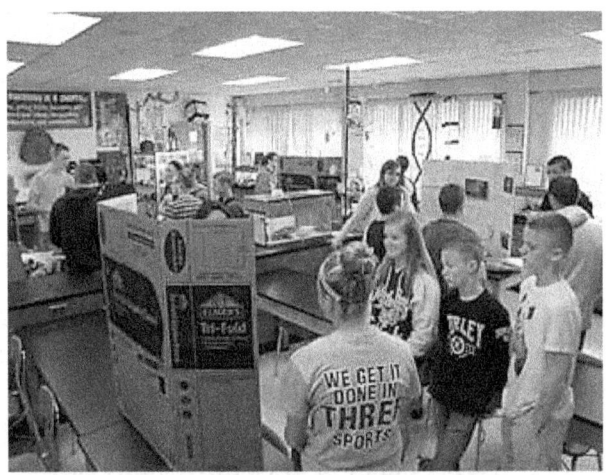

SCIENCE STRATEGIES TO INCREASE STUDENT LEARNING AND MOTIVATION IN
BIOLOGY AND LIFE SCIENCE GRADES 7 THROUGH 12

Addendum

Teaching Rubric

Trait	10–9	8–6	5–3	2–0	Totals
Eye Contact	Holds attention, seldom looking at notes during the lesson.	Consistent use of direct eye contact but returns to notes during the lesson.	Displayed minimal eye contact with audience reading mostly from notes.	No eye contact with audience as entire report is read from notes.	
Body Language	Movements seem fluid and help the audience visualize the lesson. Engagement noted.	Made movements; however, only some engagement and mingling.	Very little movement or descriptive gestures. No engagement.	No movement or descriptive gestures. No engagement or mingling.	
Poise	Displays relaxed, self-confident nature about self. Maintained professionalism.	Quickly recovers from minor mistakes. Has some tension. Some professionalism.	Displays mild tension. Has trouble recovering from mistakes. Little professionalism.	Tension and nervousness are obvious. Has trouble recovering. Little professionalism.	
Enthusiasm	Strong, positive feeling about topic during the lesson. Made lesson fun.	Occasionally shows positive feelings about topic.	Shows some negativity toward topic presented. Lesson was a chore.	Shows absolutely no interest in topic presented.	
Elocution	Clear voice and correct, precise pronunciation of terms.	Voice is clear. Pronounces most words correctly.	Voice is low. Pronounces terms poorly. Audience has difficulty hearing.	Mumbles, incorrectly pronounces terms, and speaks too quietly.	
Knowledge	Knowledgeable. Provided discussion and higher-level thinking.	Seemed a little unsure about information. Is at ease with questions.	Uncomfortable with information. Answers some questions.	Does not have grasp of information. Student cannot answer questions about subject.	

Learning Styles	Three different learning styles addressed (visual, auditory, and kinesthetic).	Two different learning styles addressed in a fair way.	One learning style addressed.	Zero learning styles were noticed.	
Mechanics	No grammatical errors. Fonts, color, background, graphics excellent.	Few grammatical errors. Fonts, color, background, graphics fair.	Some grammatical errors. Fonts, color, background, graphics all right.	Many grammatical errors. Fonts, color, background, graphics very poor.	
Manipulative	Strong, meaningful, and used appropriately. Utilized equipment in room.	Fair, somewhat meaningful, and used correctly.	Sad, not too meaningful or associated with topic.	None, failure to provide a meaningful addition to the lesson.	
Understanding	Checked for understanding at beginning, throughout, and at the end.	Checked for understanding throughout and at the end	Checked for understanding at least once during lesson.	Did not check for understanding at any time during the lesson.	

Addendum

Cancer: Nine-Page-Spread Project

A nine-page spread is a group of papers or "panels" that are closely tied or related to one another.

- The middle panel is always the "title panel," and each of the panels that surround the "title panel" are directly related to it. Each panel will include a theme.
- Each panel is to be done *professionally*, have *more than ten facts/information*, have *diagrams* (pictures, charts, and/or graphs), and have *color*. Each panel is also to depict the given *theme*. Originality and creativity will help your overall grade. Panels are to be set to *landscape* only. Try to cover the entire panel. No color printing is allowed from school; home is all right. You may have panels checked during class or wait to have the entire project done.
- Appropriate research may come from text, proper websites, study guides, and/or especially lecture notes. It is required that each panel have at least ten facts or information about the topic.

SCIENCE STRATEGIES TO INCREASE STUDENT LEARNING AND MOTIVATION IN BIOLOGY AND LIFE SCIENCE GRADES 7 THROUGH 12

- Each panel is to be put together by overlapping each panel by a half inch. Panels are to be in the same order/configuration as shown on the "Categories and Themes" chart. This may be done either inside or outside class; however, the tape must come from home.
- A copy of the rubric will be passed out in order for each member in the group to review the requirements but also to initial which panel(s) you (the student) agreed to do. Failure to do your panel will result in you (not your group) receiving a zero grade.
- Note: The tenth panel is optional; however, if done correctly, the group will receive extra credit for this assignment.

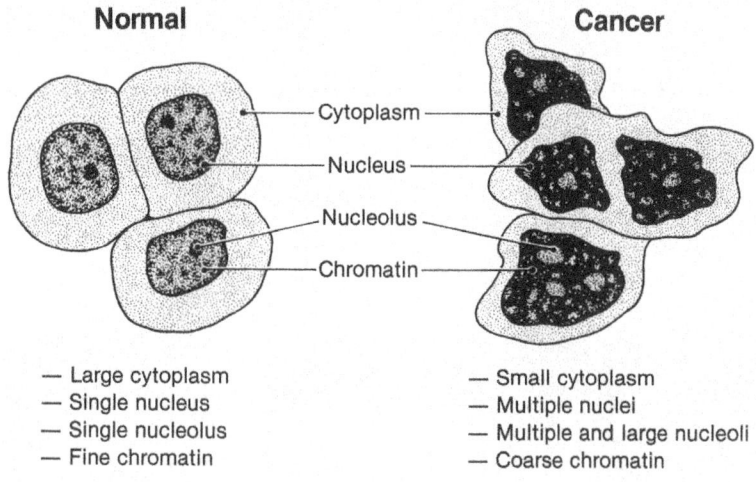

Panels: Categories and Themes

Carcinoma	Lymphoma	Sarcoma
(*Serious cartoon or comic strip, self-created* [include 10 facts/information])	(*Puzzle, i.e., crossword or word search, with 10+ questions as a key* [include 10 facts/information])	(*Foldable; ie. attach a flip book, accordion, four doors, and/ or pyramid, etc...* [include 10 facts/information])
Carcinogens	**Cancer**	**Treatments/Support**
(*Board game with 10+ questions or clues, i.e., monopoly* [include 10 facts/information])	(*Quiz with 10+ questions as a key* with everyone's name *on the top* [include 10 facts/information])	(*Advertisement, product and/or service, found in a magazine* [include 10 facts/information])

Benign Tumors (*Poem or short story, rhyming or rhymeless* [include 10 facts/information])	**Leukemia** (*VENN Diagram comparing normal and cancerous WBCs* [include 10 facts/information])	**Malignant Tumors** (*Tell how a cancer cell of your choice travels and infects a body* [include 10 facts/information])
	Cancer Ribbons (*optional*) (*Draw, color, and label 15+ different cancer ribbons*)	

Example

Addendum

BioBot

An Anatomical Collage of the Human Body

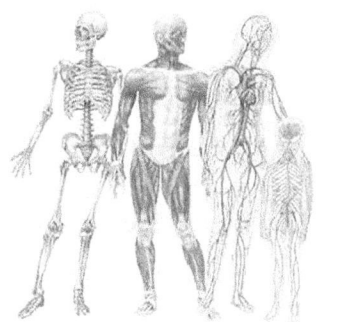
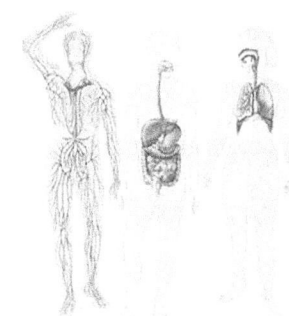

Objective

Your group is to make a BioBot that best resembles the structures and the functions of the organs and structures found in a human body.

Procedure

1. Brainstorm the best analogies for the organs and/or structures (see list) found in a human body.
   o Structure over function; however, try to accommodate both.
   o Nonperishable items only.
   o Focus on uncostly, junked, and useless items found around the home.

2. Decide who will bring materials that you'll need among the people in your group and write it down so everyone knows who will bring what.
   o Don't forget items and tools to build the BioBot.
   o Have a backup plan (i.e., phone tree / duplicate assignment responsibilities).

3. Develop a *key* that will include
   o The organs and/or structures and the item and/or analogy that you used,
   o Names of the people in your group and period.

4. Build your BioBot as soon as you begin gathering your materials.
   o Keep your area clean and organized.
   o Be imaginative yet anatomically correct.

Further Information

- All work is to be done in class (unless special permission is given).
- You may wish to bring a box, bag, or container to place parts, tools, items, etc.
- The project must have an outline or a definitive shape of the human body.
- The model is to be between one and two meters in length (no exceptions).
- Special effects are optional; however, they "may" help your grade.
- The model may be lying down or standing up.
- No items are to be used that clearly represent a structure or organ (i.e., Mr. Potato Head parts).
- Any form of poor taste during the construction or with the project will result in disciplinary action.
- Be prepared to discuss your reasonings behind your project, structures, organs, and/or terms.
- Evaluation will be done with a *project rubric*.

Required Organs and/or Structures List

brain	stomach
mouth	small intestine
eyes	large intestine
nose	rectum
ears	liver
trachea	gallbladder
bronchial tubes	pancreas
esophagus	spleen
lungs	kidneys
heart	urinary bladder
diaphragm	urethra
	ureter

Examples

Optional Organs/Structures List

skin	larynx
hair	bronchiole(s)
teeth	blood vessels
bone	appendix
tongue	ovaries *or* testis

Helping Students Be More Successful by Showcasing Work

- Rewarding
- Provides recognition
- Spotlights achievement
- Inspires future success

Strategy Examples of Showcasing Work

- Classroom walls/ceilings
- Local/school news
- Open house
- Parent-teacher conference
- School/class web page

DNA model created by two students in advanced biology

Helping Students Be More Successful by Providing Meaning

- Makes connections
- Holds students' interests
- Stimulates participation
- Gives purpose to lessons/activities
- Develops interactions

Strategy Examples for Providing Meaning

- Real-life scenarios or case studies
- Invite guest speakers
- Labs/activities (see syllabus)
- Allow students to share experiences
- Discuss syndromes/conditions
- Field trips

> **Why learn about biology?**
>
> **Because we are here to learn how the universe works.**
>
> "It is enough if one tries every day to comprehend a little of this mystery."
>
> —A. Einstein

Students working on a human evolution lab

Fruit dissection lab to compare/contrast different fruit types

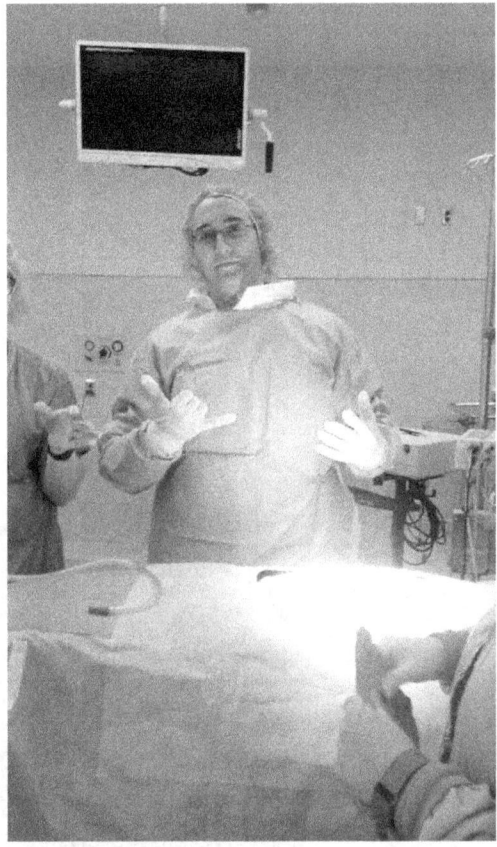

Student about to perform "simulated" heart surgery during a field trip to a local hospital

SCIENCE STRATEGIES TO INCREASE STUDENT LEARNING AND MOTIVATION IN
BIOLOGY AND LIFE SCIENCE GRADES 7 THROUGH 12

Addendum

Biology 1 (General) Lab/Activity Syllabus

Introduction to Biology

Carbon Creation Demo
- Purpose is to provide a visual of how carbon bonds to oxygen and hydrogen using sugar and sulfuric acid.

Identifying Organic Molecules Lab
- Purpose is to isolate and test major types of organic fats, proteins, and sugars using indicators.

Using a Compound Light Microscope Lab
- Purpose is to provide information about the functions and purpose of a microscope.

Sewer Lice Demo
- Purpose is to gain insight into the importance of good research and to develop proper observational skills.

Great Escape Activity
- Purpose of this exercise is to demonstrate the importance of following the scientific method.

Chinese Whispering (a.k.a. Telephone Game) Activity
- Purpose of this activity is to introduce the process of evolution using words/phrases that change or mutate from one person to another (generations).

Evolutionary Phylogenetic Cladograms and Trees Activity
- Purpose of this assignment is to illustrate how species are related via common ancestry and how characteristics help to determine relationships.

Darwin's Finches Lab
- Purpose is to use various tools to simulate different finch beaks and test these beaks on various "islands" containing different kinds of seeds and competition.

Making Cents Out of Radioactivity or the Decay Curve of Twizzlers Activity
- Purpose is to simulate the process of radioactive decay and half-life and create a graph that illustrates this process.

Ecology

Thermite Reaction Demo
- Purpose is to show what an exothermic (oxidation-reduction) reaction involves and yields using rusty ball bearings and foil.

Food Chain and Food Web Activity
- Purpose is to construct and utilize a food chain and food web using materials which include abiotic and biotic examples. Demonstrations on the effect of natural and unnatural disasters are mentioned.

Fungi Lab
- Purpose is to highlight a major contributor in the food web and in some biogeochemical cycles.

Apple Demo Activity
- Purpose is to use an apple to represent the earth and cut away sections to represent resources and populations.

Eco-Column Lab
- Purpose is for the student to create two to three ecosystems and demonstrate how they all work together.

Cytology

Cell Membrane Lab
- Purpose is to use bubbles and bubble solutions to simulate cell membranes, a fluid mosaic model, channel proteins, etc.

Zinger Bacteria Model Activity
- Purpose is to review the main structures of a prokaryotic cell during or after lecture.

Plant vs. Animal Cell Lab
- Purpose of this lab is to prepare, observe, and/or record characteristics of different types of cells such as cork, onion, human epithelial, elodea, apple peel, and/or blood cells, also including protozoa.

Shrinky Dink Cells Activity
- Purpose is to review the organelles and structures found in an animal and/or plant cell.

Hypotonic and Hypertonic Solution Lab
- Purpose is to observe osmosis across the membranes of a potato thus demonstrating various states of solutions.

The Phagocytosis and Pinocytosis Peanut Problem Activity
- Purpose of this activity is to allow the student to think about how cells ingest food by means of bulk transport by simulating the process using a peanut, string, and a clear plastic bag.

Cell Division and Reproduction

Basic DNA and Human Chromosome Model Activities
- Purpose is to introduce the basic anatomy of a DNA molecule by using eatable components that represent sugar-phosphate backbones and nitrogenous bases so students understand what DNA looks like during the interphase stage of the cell cycle. In conjunction, a life-size human model of an X chromosome will be demonstrated to represent the anatomy of what DNA looks like during mitosis and meiosis.

Mitosis Lab
- Purpose of this lab is to observe and identify the various stages of mitosis: prophase, metaphase, anaphase, and telophase. The cells that will be used will come from a plant root tip and an animal blastula. Along with studying the phases of mitosis, interphase and cytokinesis will also be reviewed.

Nine-Page Cancer Spread Project (TBA)
- Purpose is to have students review and research information discussed in class about the different aspects of cancer including carcinoma, lymphoma, sarcoma, leukemia, carcinogens, treatment/support, benign and malignant tumors, and cancer in general by making a page of each topic based on a theme and putting them all together.

Jeopardy of Something Going Wrong—Cancer Review Activity (TBA)
- Purpose is to have the student reflect on the four main types of cancer: carcinoma, sarcoma, leukemia, and lymphoma, as well as review definitions, concepts, and vocabulary about cancer by using an online Jeopardy-making puzzle (jeopardylabs.com).

Cancer Disease Pamphlet/Brochure Project Activity (*primarily an outside-class activity*) (TBA)
- Purpose is to have students pretend that they are hired by a health-care facility to develop a pamphlet or brochure for a local office. The information is to include the four main types of cancer: carcinoma, sarcoma, leukemia, and lymphoma.

Modeling Meiosis Lab
- Purpose is to use common food products and crafts to demonstrate the stages of meiosis I and II. Genetic recombination, karyokinesis, and cytokinesis will also be utilized.

Glitter: The STD of Crafts Activity
- Using glitter to represent pathogens that can cause STDs, students will be asked to begin shaking hands with two to three volunteer students to demonstrate how STDs can be spread.

Genetics

DNA Model and an Introduction to Protein Synthesis Lab
- Purpose is to create a fairly detailed model of DNA, use the model to understand the structures of the DNA, explain the location/functions of genes, and investigate the basic principles of protein synthesis.

Protein Synthesis Activity
- Using stencils to mimic genes and the classroom to represent a cell, students are introduced to the general principle of protein synthesis and the steps of transcription and translation.

DNA Extraction from Chicken Liver Cells Lab
- Purpose of this lab is to extract DNA from chicken liver cells and isolating the DNA in a solution for observation.

Sock Genetics Demo
- Purpose is to use socks as "paired genes," along with a Punnett square constructed from wood/hooks and a whiteboard, to introduce monohybrid crosses and allow the student to develop an awareness of phenotypic and genotypic patterns and fractions/percentages/ratios. Gray socks can be utilized to represent incomplete dominance and black/white socks can be utilized to represent codominance.

Making Cents Out of Punnett Squares Activity
- Purpose is to use coins and to have students that are class-led to follow an activity that demonstrates the meaning of various terms such as *genes, genotype, phenotype, dominant*, and *recessive*. This activity is also used to review the mechanics of creating monohybrid Punnett squares.

Modeling Monohybrid and Dihybrid Crosses and the Law of Probability Lab and PTC Frequency Demo
- Purpose is to predict the genotypic and phenotypic ratios of offspring resulting from the random pairing of gametes and to calculate and graph the genotypic and/or phenotypic

ratio among the offspring of a monohybrid and dihybrid cross. Colored shells, beans, and blindfolds are used to simulate the various crosses. Note: Subactivity PTC frequency, utilizing the Hardy-Weinberg law, may be demonstrated.

Human Karyotype Activity
- Purpose is to arrange drawn chromosomes in declining order of size based on height and gene band patterns in order to determine a patient's gender. A real karyotype will be presented for comparison.

Practice Pedigree Activity
- Purpose is to become aware of how a pedigree is created and used for clinical studies. Such examples as Huntington's disease and cystic fibrosis are addressed. A fun side activity will include creating a pedigree from the song "I'm My Own Grandpa."

Hardy-Weinberg Equilibrium Activity
- Purpose is to make students aware of using an expanded binomial, $(a+b)^n$, to find gene frequencies for a particular trait such as widow's peak vs. straight hairline phenotypes.

General Anatomy and Physiology

Mammalian Dissection Lab (*Note: alternative assignment available*)
- Purpose is to provide laboratory investigations into the anatomy and physiology of mammals. The lab will include a dissection of a *Rattus norvegicus* (Norway rat). External features, muscular systems, throat and oral cavity, abdominal cavity and digestive system, circulatory system, respiratory system, urinary structures, female/male genital structures, and central nervous system are covered.

Biology 2 (Anatomy and Physiology) Lab/Activity Syllabus

Histology

Anatomical Tissue Lab(s)
- Purpose is to observe various types of tissues found in the human body including human connective, muscle, nerve, epithelial, and blood tissues. Also, each tissue section of the lab will include an activity, such as model-building or tissue-naming exercises.

WebQuest: Tissue Trek Activity
- Purpose is to create a portfolio and video showing crew members encountering the five main tissues (connective, bone, muscle, nerve, and epithelial) inside a human body. The outline, description, requirements, and research are accessed online via a WebQuest website. This activity is assigned after the tissues have been discussed.

Digestive System

Finding Your Threshold of Taste Lab
- Purpose of this experiment is to determine the lowest concentration of a substance dissolved in water, which can still be tasted. The substances include sugar, salt, and vinegar.

Digestive System Dissection of the Fetal Pig Lab
- Purpose of this dissection is to expose various tissues for study. The tissues are related to the organs of the digestive system including the oral cavity, esophagus, stomach, small intestine, large intestine, rectum, liver, and pancreas.

Circulatory System

Observing Circulation Demo
- Purpose of this investigation is to observe how blood flows in a vertebrate (goldfish). The observer will be able to view blood cells traveling to and from the heart, as well as the vessels in which they travel.

Circulatory System Dissection of a Pig Heart Lab
- Purpose of this dissection is to expose heart tissues for study. Various vessels and chambers of the heart will be observed.

Contraction of the Heart Activity
- Purpose of this lab is to measure pulse rate and blood pressure and to interpret an EKG reading from human test subjects. Target heart rate may also supplement this investigation.

Making Blood Activity
- Purpose is to model and simulate blood and its components.

Human Blood Lab
- Purpose of this lab is to observe and record characteristics of different types of human blood cells.

Respiratory System

Modeling the Breathing Action Lab
- Purpose is to better understand how breathing works by constructing a working lung model.

SCIENCE STRATEGIES TO INCREASE STUDENT LEARNING AND MOTIVATION IN BIOLOGY AND LIFE SCIENCE GRADES 7 THROUGH 12

Holding Your Breath Lab
- Purpose of this lab is to test whether changes in the levels of oxygen and carbon dioxide in your blood provide the signal to stop holding your breath.

Room to Breathe Lab
- Purpose is to calculate the amount of air volume in a room and the rate in which those students in the room breathe to find out if there is enough air to breathe and/or for how long.

Urinary System

Urinary System Dissection of a Sheep Kidney Lab
- Purpose of this dissection is to expose kidney tissues for study. Various structures and sections of the kidney will be observed.

Nephron Model Lab
- Purpose of this activity is to research and construct a 3D model of a human nephron using materials from home. A brief oral explanation of how the nephron works will also be expected.

Reproductive System

Male and Female Gametes Lab
- Purpose of this microscopic investigation is to outline the structures of the human ovary and the anatomy of human sperm cells (gametes). An online search for various topics relating to human reproduction may also be implemented.

Piktochart on Teen Pregnancy Activity
- The purpose is to have students create an infographic about teen pregnancy using http://piktochart.com.

Nervous System / Senses

Presentation and Lesson Activity
- Purpose of this assignment is to present a knowledgeable explanation of the five senses and how they interact with the nervous system (spine and brain).

Thinking Cap Activity
- Students create a simple paper hat that illustrates the different parts of the brain and what those parts do.

Nervous System Dissection of a Sheep Brain Lab
- Purpose of this dissection is to expose brain tissues for study. Various sections and lobes will be observed.

Primary Anatomy and Physiology Mammalian Dissection

Making Anatomical Terminology a...Peeling Activity
- Purpose is to learn/review biological and anatomical terminology by using and dissecting a banana as a subject of study.

Felis domestica: Cat Dissection Lab
- Purpose of this lab is to summarize and investigate the various tissues and systems of a mammal (*Felis domestica*: cat) that have been discussed throughout the course. External structures, ventral muscles, dorsal muscles, thoracic and anterior systems and veins/arteries, abdominal and posterior systems and veins/arteries, urogenital systems, and related structures will be addressed.

Other(s): (May be in conjunction with any of the activities/labs/projects previously mentioned)

Anatomical Collage: BioBot Activity
- Purpose of this activity is to create a BioBot (anatomical collage) that best resembles the structures and the functions of the organs and systems found in the human body. This activity is assigned after the cat dissection is complete.

Advanced Biology 2: Dual Credit (General) Lab/Activity Syllabus

Biological Molecules

Marshmallow Molecular Models Lab
- Purpose is to use marshmallows to create 3D structures of hydrocarbons, alcohols, and lipids. Carbohydrates and proteins may also be included.

Showing the Characteristics of an Amphiphilic Molecule (Soap) Mini-Lab
- Purpose is to demonstrate the amphiphilic nature of molecules such as in soap by means of a colorful demonstration.

Preparation of Soap and the Study of Biological Fats Lab
- Purpose is to make a small batch of soap, test the soap's properties, and relate the soap to biological characteristics.

SCIENCE STRATEGIES TO INCREASE STUDENT LEARNING AND MOTIVATION IN BIOLOGY AND LIFE SCIENCE GRADES 7 THROUGH 12

G-Protein-Coupled Receptor Participation Lab
- Purpose is to demonstrate how G-proteins can influence intracellular messaging with the use of ligands by using miraculin (sour-to-sweet stimulator) and gymnemic acid (sweetness reducer).

Introduction to Entomophagy Lab
- Purpose is to investigate, cook, and consume insects as a good source of protein with little fats and carbohydrates.

Organic Breakfast Lab
- Purpose is to find and test various types of organic molecules including proteins, lipids, and carbohydrates in common food products.

Cellular Metabolism, Fermentation, and Respiration

Ghoulish Glycolysis Project
- Purpose is to review the process of glycolysis by creating a music video to describe the steps of this anaerobic process.

Fermentation Lab
- Purpose of this lab is to determine if cultured yogurt can be made by fermentation or if root beer can be made by fermentation.

Breathless about Cellular Respiration: Krebs Cycle Lab
- Purpose is to study how exercise affects the disposal of waste (CO_2) from cellular respiration.

Photosynthesis

Chloroplast in Guard Cells Mini-Lab
- Purpose is to simply identify the chloroplast organelles, where photosynthesis takes place, and the guard cells that help regulate the materials necessary for photosynthesis.

Plant Pigments for Photosynthesis Lab
- Purpose is to have pigments extracted from various leaves and then see if the solvent in which the pigments are exposed to will separate those various pigments (i.e., carotenes, xanthophylls, chlorophylls a and b), paper chromatography activity.

Exciting Electrons that Have Nowhere to Go Activity
- Purpose is to demonstrate florescence when photons of light excite electrons to a higher energy level and not be received by an electron acceptor molecule such as $NADP^+$.

Measuring the Rate of Photosynthesis Lab
- Purpose of this investigation is to observe photosynthesis in an aquatic plant and measure the rate of photosynthesis-based absorption of different colors provided to the plant.

Photosynthetic Production of Starch Lab
- Purpose is to extract and indicate starch from sugar-producing leaves.

DNA, RNA, and Protein Synthesis

DNA Origami Activity
- Purpose is to create and model a DNA from paper.

Extracting DNA for Human Cells Lab
- Purpose of this lab is to extract DNA from human cheek cells and isolating the DNA in solution for observation.

Building a Five-Plus-Meter DNA Model (*optional extra credit*)
- Purpose is to create a representation of a DNA molecule in order to review the DNA's anatomy and function.

Two-Ply DNA Activity
- Purpose is to use common household paper products to simulate DNA replication.

Protein Synthesis Modeling Lab
- Purpose is to create an mRNA from a DNA sense strand (via transcription) in order to make a protein (via translation) using crafts.

DNA, RNA, and Protein Synthesis Activity and Computer Simulations Online Activity
- Purpose of this activity is to illustrate how the order of nucleotides in DNA determines the order of amino acids in proteins. It is used to reinforce the concept that any change in the order of nucleotides can change the order of amino acids in proteins (internet-based activity).

Genetic Expressions

The Biology Project Online Activity
- Purpose is to conduct activities online that address such topics as Mendelian genetics, karyotyping, sex-linked inheritance.

SCIENCE STRATEGIES TO INCREASE STUDENT LEARNING AND MOTIVATION IN BIOLOGY AND LIFE SCIENCE GRADES 7 THROUGH 12

BLAST Online Activity
- Purpose is to use the website from the National Center for Biotechnology Information (NCBI), www.ncbi.nlm.nih.gov, to identify a sequence of bases from a DNA sample. Sequence numbers, official names, gene loci, and phenotypic descriptions are addressed.

Human Karyotype Paper and/or Online Activity
- Purpose is to arrange chromosomes in declining order of size based on height and gene band patterns in order to determine a patient's gender and disorders such as Klinefelter's or Turner's syndrome.

Pedigree Activity
- Purpose is to become aware of how a pedigree is created and used for clinical studies. Such examples as congenital ptosis, Tay-Sachs, hypercholesterolemia, and muscular dystrophy are addressed.

Stickleback Gene Activity
- Purpose of this activity is to demonstrate how gene switches (and those that are mutated [i.e., pelvis gene]) are used to turn on genes for various phenotypic characteristics between fresh and saltwater stickleback fish (i.e., pelvis, jaw, nose).

RNAi Activity
- Purpose is to use two analogies in the activity to illustrate how RNAi's are used to "interfere" with specific gene expression and protein production.

Patterns and Gene Expression Presentation Using Presentation Software (TBA)
- Purpose is to create a presentation of ideas, diagrams, definitions, information, and/or examples based on genetic expression such as gender linkage, nondisjunction of chromosomes, gene mutations, human genome project, gender determinations, epigenetics, and RNAi.

Plant and Animal Diversities

Presentation and Lesson Activity
- Purpose of this assignment is to present a knowledgeable explanation of one of the five main biomes and how plants and animals interact with one another in nature via PowerPoint and/or Prezi (or others).

Anatomical and Physiological Diversities Lab
- Purpose is to observe plant (dicot) and animal (chicken) embryos in order to demonstrate diversity among organisms.

Plant and Animal Microhabitat and Ecosystem Lab (TBA)
- Purpose is to research and create an APA-style lab that studies single and multicellular plants and animals in their environment.

Evolution

Presentation and Lesson Using EVO Activity
- Purpose of this assignment is to present a knowledgeable explanation of evolution. Various lessons and videos will accompany this activity including hominid skull models.

The Amazing Human Race Activity
- Purpose of this assignment is to provide students the opportunity to examine models of hominids from millions of years of evolution by making comparative observations and developing hypothetical conclusions about human ancestry.

Biology 2 (Botany) First Semester Lab/Activity Syllabus

Introduction to Botany and Plant Diversity

Chia Sculpture Lab
- Purpose is to provide a metaphorical "hook" to gain an initial interest in botany by making, sculpturing, and nurturing a self-made chia sculpture.

Anatomy and Function of *Polytrichum* (Moss) and Leafy Liverwort Lab
- Purpose is to examine moss and liverwort in order to help understand the anatomy and physiology of bryophytes.

Anatomy and Function of *Pterophyta* (Ferns) Lab
- Purpose is to examine a fern in order to help understand the anatomy and physiology of tracheophytes.

Dissection of a Dicot Seed Lab
- Purpose is to examine the anatomical structures of a dicot seed and embryo.

Seed Germination and Plant Development Lab (Note: Continues into "stems, roots, and leaves.")
- Purpose is to grow and observe plant development from seed to near maturity using a controlled experiment and to make daily observations, records, and analyses for the growing plant. An APA research paper based on the results will be formulated.

SCIENCE STRATEGIES TO INCREASE STUDENT LEARNING AND MOTIVATION IN BIOLOGY AND LIFE SCIENCE GRADES 7 THROUGH 12

Stems, Roots, Leaves

Ground Tissue Salad Demo
- Purpose is to make students aware of various types of ground tissue in common foods.

Plant Tissue, Anatomy, and Physiology: Roots Lab
- Purpose is to observe and understand the external and internal structures and functions of roots.

Plant Tissue, Anatomy, and Physiology: Stems Lab
- Purpose is to observe and understand the external and internal structures and functions of stems.

Plant Tissue, Anatomy, and Physiology: Leaves Lab
- Purpose is to observe and understand the external and internal structures and functions of leaves.

Reproduction and Development of Seed Plants

Angiosperm Dissection: Flower Lab
- Purpose is to identify structures found in a "complete" flower. The reproductive structures will be primarily targeted during this investigation.

Angiosperm Dissection: Fruit (Fleshy Variety) Lab
- Purpose is to identify structures and tissues found in "fleshy" fruits.

Seed Respiration Lab
- Purpose is to measure the consumption of oxygen of a germinating seed by creating and using a respirometer.

Reduction of Starches in Banana Cells Lab
- Purpose of this activity is to compare the amount of starch in unripe vs. ripe bananas in order to provide reasoning as to why banana tissue softens and becomes sweeter when they ripen.

Other(s): (Maybe in conjunction with any of the activities/labs/projects previously mentioned)
Supermarket Botany Online Activity
- Purpose is to relate common foods found in a grocery store to studied plants. The online site is http://www.csu.edu.au/research/grahamcentre/education/Supermarket%20Botany/ ie5/index.html?dhtmlAtivation=inplace.

Dissecting Microwave Popcorn Activity
- The purpose is to discover the unique design of microwave popcorn packages and to investigate brands, kinds, and percentage of corn popped.

Biology 2 (Zoology) Second Semester Lab/Activity Syllabus

Introduction to Zoology

Introduction to Animal Symmetry Activity
- Purpose is to distinguish bilateral and radial symmetry with the use of clay animals created by the students themselves.

Sponges and Cnidarians

Sponge "Cake" Round "Porifera" Activity
- Purpose is to create a model of a typical sponge from sponge cake.

Investigating Hydras (Green and/or Brown) Lab
- Purpose is to observe the anatomy of hydras. This investigation will allow one to explain the behavior of hydras as well.

Worms and Mollusks

Planarian (Worm) Lab
- Purpose is to observe a flatworm's anatomy, behavior, and asexual reproductive ability.

Dissection of an Earthworm (Worm) Lab
- Purpose of this dissection is to expose tissues and organs for study. Various characteristics and terms will be observed and addressed.

Dissection of a Clam (Mollusk) Lab
- Purpose of this dissection is to expose tissues and organs for study. Various characteristics and terms will be observed and addressed.

Arthropods and Echinoderms

Dissection of a Crayfish (Arthropod) Lab
- Purpose of this dissection is to expose tissues and organs for study. Various characteristics and terms will be observed and addressed.

SCIENCE STRATEGIES TO INCREASE STUDENT LEARNING AND MOTIVATION IN BIOLOGY AND LIFE SCIENCE GRADES 7 THROUGH 12

Dissection of a Grasshopper (Arthropod) Lab
- Purpose of this dissection is to expose tissues and organs for study. Various characteristics and terms will be observed and addressed.

Dissection of a Starfish (Echinoderm) Lab
- Purpose of this dissection is to expose tissues and organs for study. Various characteristics and terms will be observed and addressed.

Non-vertebrate Chordates; Vertebrate Chordates: Fishes and Amphibians

Dissection of a Perch (Fish) Lab
- Purpose of this dissection is to expose tissues and organs for study. Various characteristics and terms will be observed and addressed.

Dissection of a Grass Frog (Amphibian) Lab
- Purpose of this dissection is to expose tissues and organs for study. Various characteristics and terms will be observed and addressed.

Reptiles and Birds

Dissection of a Turtle (Reptile) Lab
- Purpose of the dissection is to expose tissues and organs for study. Various characteristics and terms will be observed and addressed.

Dissection of a Pigeon (Bird) Lab
- Purpose of this dissection is to expose tissues and organs for study. Various characteristics and terms will be observed and addressed.

Mammals (TBA)

Blubber Glove Lab
- Purpose is to simulate blubber which helps to explain endothermic characteristics of mammals.

Hair, Hair, Hair Lab
- Purpose is to compare and contrast different hair follicles from various mammals under the microscope.

Mother's Milk Emulsification Demonstration
- Purpose is to provide visual evidence of the emulsification properties of soap and provide evidence that milk contains fat.

Using Owl Pellets to Study Mammalian Remains Lab
- Purpose is to view and hypothesize what type of mammal was regurgitated by a raptor to study skeletal and anatomical features of mammals.

Helping Students Be More Successful by Keeping It Professional

- Gains respect
- Garners class reputation
- Shows pride
- Demonstrates success
- Becomes a role model

Strategy Examples for Keeping It Professional

- Salutations and dress
- Ethical and trustworthy
- Grammar and language / polite yet firm
- Punctual, organized, prepared, flexible, devoted
- Use common sense and sound judgment
- Take ownership of your work and find a style of teaching that works for you
- Admit defeat—It's okay…you're only human.
- Be supportive of parents and administration (see focusing on biology—first-day letter).
- Demonstrate validity (i.e., topics corresponding to goals and standards) and reliability (i.e., exams that measure what they are supposed to) when creating quizzes/tests.
- Include a variety of properly written questions for exams (see introduction to biology, tools, and evolution chapter exam example).
- Take pleasure in successfully surviving the day but want to do it again the next day.

Can different styles of teaching sometimes be evident in the clothes we wear (i.e., Stripe Day at school)?

- Avoid the "Curse of Knowledge":
  - We do not remember what it is like to *not* know what we are trying to teach.
  - We cannot relive the difficult process of learning our content.
  - We end up assuming our lesson's content is easy, clear, and straightforward. We assume that connections are apparent and made effortlessly (example: Better to do Banana autopsy mini-lab prior to a dissection).

Addendum

Making Anatomical Terminology a…Peeling A "Bamana/Bawomana" Autopsy Activity

Some basic biological terminology should be studied and understood when doing anatomy. By doing this activity, you should become more familiar with terms that are used when referencing body parts and positions. Note in this activity, the stem of the banana is the top and the curvature would represent the back as it is facing you.

Procedure

Materials

Banana	Scalpel/Knife
Sharpie	Paper Towel
Ruler	References

Steps

Making Your "Bamana/Bawomana"—*Don't slip up now!*

1. Lay your *bahumana* in its anatomical position with the *ventral* (belly) side facing you.
2. Use a Sharpie and draw eyes and a mouth on the *anterior* (front) *facial* (face) *region* and a belly button on the *abdominal* (belly) *region*.
3. Turn the *bahumana* around to the *posterior side* (back side) so that side is facing you.
4. Draw two circles on the *dorsal* (back) *inferior* (below) *region* toward the *posterior/caudal* area to represent "butt cheeks."
5. On the *lateral* (toward the side) *sides* of the specimen in the correct *superior* (above) *cranial* (head) *region*, draw ears onto your *bahumana*.
6. Draw arms and legs on the *lateral* (toward the side) *portions* onto your *bahumana*.

Check Point: Verbally identify the following with your instructor before moving on.

__ ventral	__ abdominal	__ lateral	__ caudal
__ anterior	__ posterior	__ superior	__ dorsal
__ facial	__ inferior	__ cranial	

Doing an External Physical—*I know. You might not be peeling too well after this!*

7. Do an external physical exam of your *bahumana* by looking for abnormalities, such as wounds, scars, disfigurements, colorations, etc. With the bullets provided, list two of these descriptions; however, you *must use* anatomical directions/positions from table 1 to do so. *For example, near the bahumana's anterior cranial regions, there are two bruises or brown spots three milliliters in diameter.*

Check Point: Verbally show these descriptions with your instructor before moving on:

Doing an Internal Physical—*Time for this banana to split!*

8. Return your *bahumana* to its anatomical position.
9. Carefully make a *transverse* (cross section) cut about five centimeters *inferior* (toward the feet) to the top of the "head" at the *cervical* (near the neck) region but above the arms of your specimen. *Hint: Essentially, you are decapitating your bahumana.*
10. Make another *transverse* (cross section) cut at the *umbilical* (belly button) region but above the legs. *Hint: You should have three pieces.*

Check Point: Show the two longitudinal views to your instructor before moving on.

11. Using the *cranial* (head) *region* created from step 9, carefully make a *frontal* (coronal) incision all the way through. The result of this cut is now referred to as a longitudinal view.
12. Using the *trunk* (mid) region created from step 9, carefully make a *midsagittal* (down middle/center separating right and left sides equally) incision all the way through. Note: A *parasagittal* plane would produce unequal halves.
13. Make an *oblique* (plane at an angle/diagonal) section through the section containing the legs.

Check Point:

- Show the results of steps 11–13 by placing your results on a paper towel and labeling them the following: Frontal Incision, Midsagittal Incision, and Oblique Incision.
- Answer the following by *circling* the correct response:
    o Is the "face" of the *bahumana* found caudal *or* cranial to the specimen?
    o Is the "hand" of the *bahumana* found proximal *or* distal to the specimen's chest?
    o Is the "back" of the *bahumana* found ventral *or* dorsal to the specimen?
    o Is the "feet" of the *bahumana* found superior *or* inferior of the specimen?

Show labeling and answers to the questions to your instructor. Please clean up and return all materials as directed.

Table 1

Directions or Positions

Term	Definition	Term	Definition	Example
Anterior	toward the front	*Posterior*	toward the back	The stomach is anterior to the kidneys.
Ventral (used primarily in quadrupeds)	belly side	*Dorsal* (used primarily in quadrupeds)	backside	The vertebrae are dorsal to the stomach.
Medial	toward the midline of the body	*Lateral*	toward the side of the body	The eyes are lateral to the nose.
Superior	toward the head	*Inferior*	toward the feet	The hips are inferior to the shoulders.
Cranial or Craniad	toward the head	*Caudal or Caudad*	toward the tail	The skull is craniad to the vertebrae.
Superficial	toward the surface	*Deep*	toward the core or center	The skin is superficial to the muscles.
Proximal (usually refers to the appendages)	closer to the body	*Distal* (usually refers to the appendages)	farther from the body	The elbow is distal to the shoulder and proximal to wrist.
Palmar	the palm of the hand	*Dorsal*	the back of the hand	Fingerprints are located on the palmar surface.
Plantar	the sole of the foot	*Dorsal*	the "top" of the foot	The d. pedis artery is on the dorsal surface of the foot.

Planes or Sections through the Body
Transverse (Cross Section)—Perpendicular to the long axis of the body
Sagittal—A longitudinal section separating the body into right and left sides
Frontal (Coronal)—A longitudinal section dividing the specimen into dorsal and ventral parts

Addendum

Focusing on Biology

Dear Parent(s) and/or Guardian(s),

The main goal of this letter is to establish communication between you and myself. Your student will get information about several items pertaining to the class, including objectives, materials/supplies, curricula, agendas, policies, grading, studying suggestions, tutoring, and other general information.

I would like to invite you to review the information and policies that I will be sharing with the class (handout and via the classroom website) and to invite you to contact me (phone or e-mail) if you have any concerns and/or questions. I would also like to take this opportunity to remind you to continue encouraging, helping, and supporting your student *throughout* the year. My goals, objectives, and work ethics are very high, and I expect the same from my students. Help your student by being involved in their education (checking assignments, getting organized, and asking about their progress). Students can always ask for help by asking questions in class, stopping by outside class, by checking the classroom website, or by using my e-mail. For their convenience, I have also posted a "tutoring schedule" outside my door showing when I am available.

Communication and awareness of your student's progress is very important.

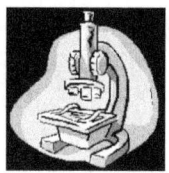

- Online gradebook to view your student's progress often.
- Class website to view student work, lesson plans, tutorials, policies, etc.
- Lesson plan outline to view class topics, assignments, activities, labs, exams, etc.
- Google drive to view and access most of the assignments, activities, labs, tutorials, etc.
- E-mail and phone to contact you if necessary (have them registered and correctly updated at the school office)
- www.remind.com to receive quick, one-way, and private messages only from me to keep you updated.

Three ways to sign up for free (optional for parents/guardians but highly recommended):

*   Text your information.
*   E-mail your information.
*   Visit www.remind.com, then click "I'm a Student/Parent" entering the class code (xxxxxx) then choose phone or e-mail. (You may need to check junk or spam folder. To learn more or unsubscribe, go to www.reminder.com.) Note: Each student will be highly encouraged to sign up for their own Remind account as well based on their class.

All eleventh- to twelfth-grade students may *only* take *dual credit* in anatomy/physiology, advanced biology, or botany/zoology class, though Ball State University will be notified about how to sign up for college credit early in the year. Be aware that *all* information and billing provided electronically or on paper will be addressed to your student *only*.

All ninth- to twelfth-grade students who took or will be taking biology this year are encouraged to join our *Biology Club*. The club focuses primarily on helping local fundraisers, science-related activities outside school, and utilizing social media to communicate ideas and information about biology. All twelfth-grade students who participate accordingly will get to wear and keep the Biology Club graduation cord. Notification about the club will be made early in the year, a great opportunity to get involved outside the classroom.

Lastly, it is my policy to start class right away, so your student needs to prepare a few key materials within a day or two of class:

>   Composition notebook (*only*)
>   Accordion binder (*only*)
>   Pencil/pen
>   Colored pencils (*24+ high quality*)
>   Loose leaf paper
>   Calculator
>   Online text

You can contact me by

>   Phone:
>   E-mail:
>   Web:

I look forward to a great and informative year.

Mr. D. Butler

SCIENCE STRATEGIES TO INCREASE STUDENT LEARNING AND MOTIVATION IN
BIOLOGY AND LIFE SCIENCE GRADES 7 THROUGH 12

Addendum

Introduction to Biology, Tools, and Evolution Exam

True or False: Place a *T* in the blank if the statement is true or an *F* if the statement is false (2 points each).

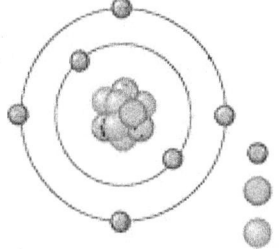

1. Protons, neutrons, and electrons make up atoms.
2. The procedure or experiment of a lab includes three main categories: materials and steps and results.
3. An isotope can be several atoms of the same type.
4. Evolution can basically be defined as "change."
5. A compound is made from nonmetal or metal elements bonded together.
6. A suitable hypothesis: if bacteria on skin are exposed to warm and wet climates, then they may multiply in number.
7. When making observations, make an effort to record everything, use the right tools, and get involved.
8. Common descent refers to the fact that all organisms on earth are related.
9. The following are the *only* characteristics of life we know: movement, energy, adaptation, growth, excretion, response, water, organic, reproduction, and metabolism.
10. Groups of molecules can make organelles, such as the nucleus, ribosome, and mitochondrion.
11. Cells are made up of organelles.
12. If cells communicate with one another, they can form skin, muscle, or bone.
13. It's best to research and gather background information about a problem before stating a hypothesis.
14. Organs that perform mainly together for a single function is called a system.
15. Protozoa, despite not having systems, such as an amoeba or paramecium are still considered to be organisms because they exist and survive independently.

Matching: Match the best term with its definition (2 points each).

16. The study of life
17. Based on facts and not just a guess or hunch
18. A group of cells performing together
19. Proportion of elements together (i.e., $C_gH_{12}O_6$ → 1:2:1)
20. An educated statement that can be tested

A. Theory
B. Molecules
C. Biology
D. Tissues
E. Hypothesis

Matching: Place the term with the best definition or example when classifying humans (2 points each).

Classification   Choices

21. Kingdom   A. Primate
22. Phylum   B. Chordata
23. Class   C. Hominid
24. Order   D. Animalia
25. Family   E. Mammalia

Multiple Choice: Choose the best response to the following statements and/or questions (2 points each).

26. The idea that individuals whose characteristics are suited to their environment will survive.
    A. evolution   C. spontaneous generation
    B. hypothesis   D. natural selection

27. This sends a beam of electrons *across* a specimen.
    A. stereo microscope   C. simple microscope
    B. compound microscope   D. scanning electron microscope

28. May have more than two lenses; however, when used properly, you are only looking in two lenses.
    A. stereo microscope   C. simple microscope
    B. compound microscope   D. scanning electron microscope

29. An example of humans influencing natural selection would *not* be
    A. peppered moth   C. resistant bacteria
    B. tolerable pesticides   D. Galápagos finches

SCIENCE STRATEGIES TO INCREASE STUDENT LEARNING AND MOTIVATION IN BIOLOGY AND LIFE SCIENCE GRADES 7 THROUGH 12

30. In regard to its wavelength, the electron microscope shows more detail because the wavelength in which an electron "rides"
    A. is shorter than light.
    B. covers more area on a specimen.
    C. A and B are correct.
    D. is longer than light.

31. The following would be a specific example of just several species living together.
    A. community
    B. ecosystem
    C. population
    D. biosphere

32. Which of the following provides evidence for evolution?
    A. fossils and convergence
    B. anatomy and transition life-forms
    C. embryology and genetics
    D. all are correct

33. Change in traits that occur mainly in small populations caused by randomness (i.e., may be American Indians blood types)
    A. genetic drift
    B. natural selection
    C. evolution
    D. biome

34. Created the first practical yet simple microscope
    A. Zach and Hans Janssen
    B. Francis Bacon
    C. Anton van Leeuwenhoek
    D. Louis Pasteur

35. The ocular lens is 10×, and the objective lens is 40×. What is the total magnification (not resolution) of the scope?
    A. 10×
    B. 50×
    C. 400×
    D. not enough information

36. According to evolution and science, which of the following is *not* true?
    A. Humans share a common ancestor with modern-day monkeys and apes.
    B. Humans did not live with dinosaurs; however, ancestral mammals did.
    C. Humans evolved in a branching (treelike) format on an earth that is billions of years old.
    D. Humans came from monkeys, lived with dinosaurs, and evolved in a linear pipeline manner.

37. Nonliving items, as well as living organisms, have _____, which are made up of protons, neutrons, and electrons.
    A. cells
    B. organelles
    C. atoms
    D. systems

38. Author of the book *On the Origin of Species*—this celebrated individual explained how natural selection and scientific evidence describe the process of evolution.
    A. Lazzaro Spallanzani
    B. Francesco Redi
    C. Louis Pasteur
    D. Charles Darwin

39. Proved that maggots do not spontaneously appear from rotten meat, even with air, by using a screened topped jar full of meat.
    A. Alfred Wallace
    B. Francesco Redi
    C. John Needham
    D. Louis Pasteur

40. Proved that microorganisms would not grow in a sealed container by melting the top of his flask to prevent air from getting in.
    A. Lazzaro Spallanzani
    B. Francesco Redi
    C. John Needham
    D. Louis Pasteur

41. Helped prove that even with air, a type of spontaneous generation involving microbes from broth would not occur.
    A. Lazzaro Spallanzani
    B. Alfred Wallace
    C. John Needham
    D. Louis Pasteur

42. Can specifically increase extinction because of inbreeding or genetic drift (i.e., causing Ellis-van Creveld syndrome).
    A. kinship similarities
    B. subtlety of evolution
    C. natural selection
    D. subpopulation isolation

43. Accepted that a new species develops from the passing of traits to offspring that allows an organism to survive.
    A. Francesco Redi
    B. John Needham
    C. Charles Darwin
    D. Louis Pasteur

44. Can, for example, separate white blood cells from red blood cells in a solution of blood.
   A. stain
   B. centrifuge
   C. micromanipulation
   D. vital stain

45. Evolution is based on scientific facts, confirmed predictions, and testable hypotheses, including successful genetic mutations and _____?
   A. ancient fairy tales
   B. natural selection
   C. preconceived notions
   D. spontaneous generation

46. Which of the following are important elements found in the cells of biological organisms?
   A. Hydrogen, Carbon, Sulfur
   B. Carbon, Nitrogen, Hydrogen
   C. Phosphorus, Oxygen, Nitrogen
   D. All are important, including Calcium and Iron.

47–55. **Labeling a Compound Microscope** (2 points each)

Label the parts of a microscope by filling in the blanks with the matching number on your answer sheet (see term choices below).

Choices

Arm
Light
Fine Focus Knob
Ocular
Diaphragm
Coarse Focus Knob
Base
Stage
Objective

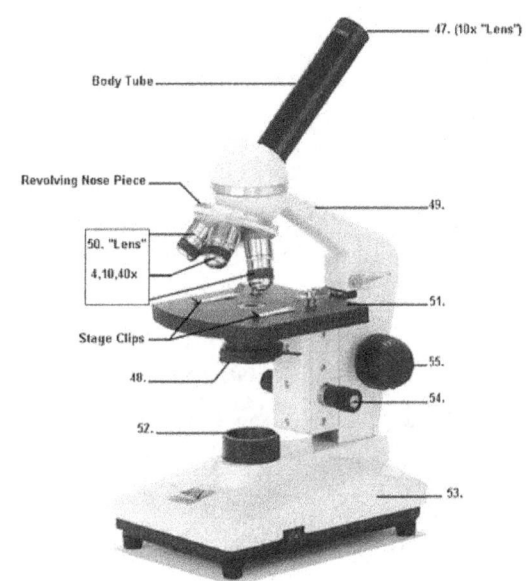

Pasteur's Flask (10 points possible)

56. Answer the following question on the *top-back-half* side of your answer sheet:

Louis Pasteur was credited with solving the spontaneous generation mystery by creating an instrument that proved organisms would not grow in broth even with air.

Question

Explain in detail how microbes didn't come from broth by *drawing* the special flask used by Pasteur *and* by *labeling* the drawing with the following terms: *Broth, Dust Particles, Swan or Gooseneck, Microbes, No Microbes, and Air or Airflow.*

Darwin's Discoveries (10 points possible)

57. Answer the following question on the *bottom-back-half* side of your answer sheet:

Evolution was publicized by a gentleman named Charles Darwin who, with the help of another naturalist named Alfred Wallace, developed the understanding of evolution based on geological and biological evidence, natural selection of species, and beneficial mutations. Thus, scientists are now able to deduce that successful organisms exist today after millions of years because they were able to pass on their favorable traits to their offspring.

Question

Choose only one (1) of the following examples from the lecture:

- Tortoises
- Finches
- Iguanas
- Prickly pear

Number and *answer each* of the following statements and questions below from your example selected above as short, complete, yet detailed responses in order to explain an evolutionary characteristic based on an apparent beneficial mutation trait and natural selection:

1. Name the beneficial mutation that was discussed in class for your chosen organism.
2. Name the geological place where Darwin noticed the organism and its characteristics.
3. Tell how the organism is able to adapt to its environment based on the mutation.
4. Explain why natural selection will allow the organism to pass its mutation to its offspring.
5. What evidence could you state about your organism to provide facts about its evolution?

*Diagram: Draw a picture to go along with your explanations (optional).

Helping Students Be More Successful by Having an Open-Door Policy

- Opportunity for students
- Validates a willingness to help
- One-to-one benefits
- Positive teacher-student relationship
- Quick chat or verifications

Strategy Examples for an Open-Door Policy

- Business cards or pamphlets / first-day pencil gifts
- School e-mail accessibility
- Be flexible when scheduling meetings
- Have your door physically open
- Invite parents/guardians
- Be willing to stop what you're doing
- Post class and tutorial schedule physically (outside the door) and virtually (in cyberspace)

Biology... the Study of Life
ie. Your Website Here

* Lesson Plan Outlines and Grades
* Curriculum, Standards, Policies
* Resources, Tutorials, Study Ideas
* Student Work and Extra Credit
* Setting Up eText and REMIND
* Biology Club Sign-Up...and More

You are always welcome to contact me any time for questions, concerns, study suggestions, etc...

Email:
Phone:

See Website or Tutorial Schedule for Suggested Days and Times to Meet; Otherwise Door is Always Open

Ideas to Do Well in this Class
* Avoid Being Absent
* Try a Memory Technique
* Check Your Grades Often
* Be Prepared Before Class
* Understand the Vocabulary
* Manage Your Time Wisely
* Try Teaching Others the Topic
* Attend Study Sessions if Offered
* Do All the Work and Assignments
* Take Great Notes During Lecture
* Study with No Distractions/Timed
* Focus on Unfamiliar Concepts Only
* Ask Questions/Seek Help if Needed

Front and back business cards given to students during the first week of school and parents during conferences, along with the proceeding biology brochure/pamphlet.

Addendum

Biology Club

Open to all biology students, past or present, willing to truly participate and be active in science and the community.

See classroom website, SW Biology Club website, or SW Biology Club Facebook page for more information.

The mission of the Southern Wells Biology Club is to recruit, support, nurture, and promote students with an interest in biological sciences for personal reasons, academic preparation, the betterment of society, and career opportunities by providing guidance, resources, and activities to meet these goals.

Like us on Facebook!

Course Objectives

To develop a foundation in biological vocabulary and principles
To develop an awareness about the uniqueness and diversity of life
To develop an appreciation of the interrelationships among living organisms
To develop an understanding of some natural laws and their applications to life
To develop the concept of commonality of structure and function in living organisms

Axioms of Biology (Starting Points of Reason)

Cells
Heredity
Evolution
Energy
Regulation

Biology

The Study of Life

Classroom website

Class information, lesson plans, policies, curricula, syllabi, study strategies, grades, extra credit, sign-up directions, research, student works, and much more!

Addendum

General Study Hints and Guides

1. Don't be absent from class, and be attentive while in it. Attendance is very important because it allows you to stay on top of assignments, information, lectures, activities, exams, projects, etc. Get good night's rests, and eating healthy helps. Ask questions, and participating in class will keep the lesson more interesting.
2. Be prepared before going to class. Have all the necessary materials for class, such as pencils/pens, paper, folders, texts, and assignments. Try your best to stay organized.
3. Correctly do the classwork that is assigned to you. Keeping up with daily reading and work assignments will prevent you from falling behind and having to constantly be catching up. By doing the work, it will keep you on your toes in the event that the teacher asks you a question or gives a quiz. Plus, this allows you to ask the teacher questions if something is unclear instead of waiting the night before a test when you wouldn't have the opportunity

SCIENCE STRATEGIES TO INCREASE STUDENT LEARNING AND MOTIVATION IN BIOLOGY AND LIFE SCIENCE GRADES 7 THROUGH 12

to ask the question(s). Also, understand that not everything is graded. The works you do are to help you in the long run (i.e., chapter exam).

4. Write it all down during lecture. Take good notes (never recopy), make note cards, write down questions, and ask them as soon as possible. If you doodle in your notes, it's all right. The picture might remind you of a concept during an exam.

5. Attend study sessions if they are offered (either class sessions or classmate sessions). You may have forgotten something, or you may have a question that someone at the session can answer.

6. Communicate with your teacher. For whatever reason you're not sure of a concept, assignment, lesson, etc., go talk to your teacher. Schedule a time to meet for tutoring. Teachers are happy to meet your needs if you're sincere and independently responsible for your studies. Remember, the teacher generally teaches up to twenty-five brains in the classroom. If they can focus on just one brain (yours), the process of helping is a lot easier for both participants.

7. Manage your time. For example, even though an assignment was not given in class, take some time (ten minutes, twenty minutes, thirty minutes, etc.) to review what you went over that day. This will generate questions that you can ask the teacher the next day and/or make you more familiar with the material so that studying for the test will not be a totally new experience. Don't procrastinate and get assignments done. Study for the test early. If you have any questions, ask them now. Study *nearly* every day. Don't wait until the night of.

8. Know the vocabulary to speak the topic. A lot of times, certain subjects are like (or are) foreign languages. In order to speak the language, you have to know the terms. Daily review, relating personal experiences, or the use of analogies can help.

9. Try teaching others. Sometimes you learn if you teach it. While studying for a quiz or test, literally try teaching the topic to a friend or family member. Discuss with them what you've learned and even try asking them questions. Review the answers with them while looking at your notes or texts. In other words, you be the teacher. This can help reduce anxiety or embarrassment because you've been in the class, and you have all the answers.

10. Don't try to multitask. Being distracted by TV, cell phones, electronic games, lyrical music, internet, etc. can actually cause you not to concentrate and focus on your studies. Find a quiet area to study without being bothered by outside influences. After you study, then reward yourself with music, TV, texting, internet, etc.

11. Only study items that you don't understand. There is really no need to study items or concepts that you're already familiar with. It takes extra time and can give you a false sense of security that you know all the information. Concentrate on what you don't know.

12. Try memory techniques. Use mnemonics, chunking, or the "memory palace" (see Google Drive or class website for details).

Note: These are only suggestions. See Google Drive or class website or stop in for more ideas until we find one that works (Mr. Butler).

Addendum

Class and Tutorial Schedule

(*Subject to Change*)

* If possible, schedule appointments first (i.e., before or after school), but this is not truly necessary.
* Get all necessary passes from other classes or coaches to visit during school and practice hours.
* Tutoring/help is mainly available before/after school, during prep, homeroom, and/or supervision times.
* Best to have specific questions or concerns prior to appointment.
* Typically, I am at school between 7:15 a.m. and 3:05 p.m.

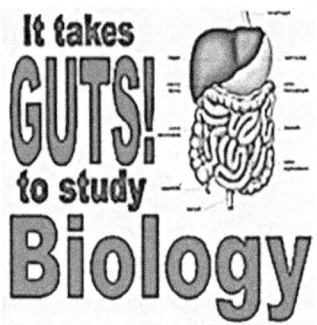

Time	Monday	Tuesday	Wednesday	Thursday	Friday
7:15–7:45	*Tutoring/ Help Time*	*Tutoring/ Help Time*	*Tutoring/ Help Time*	*Tutoring/ Help Time*	*Tutoring/ Help Time*
Period 1	Anatomy/ Physiology	Anatomy/ Physiology	Anatomy/ Physiology	Anatomy/ Physiology	Anatomy/ Physiology
Period 2	*Prep Tutoring/ Help Time*	*Prep Tutoring/ Help Time*	*Prep Tutoring/ Help Time*	*Prep Tutoring/ Help Time*	*Prep Tutoring/ Help Time*
Period 3	General Biology	General Biology	General Biology	General Biology	General Biology
Period 4	Botany/Zoology	Botany/Zoology	Botany/Zoology	Botany/Zoology	Botany/Zoology
Period 5	Advanced Biology	Advanced Biology	Advanced Biology	Advanced Biology	Advanced Biology
Period 6	General Biology	General Biology	General Biology	General Biology	General Biology
Period 7	General Biology	General Biology	General Biology	General Biology	General Biology
3:10–3:30+	*Tutoring/ Help Time*	*Tutoring/ Help Time*	*Tutoring/ Help Time*	*Tutoring/ Help Time*	*Tutoring/ Help Time*

You can also visit the classroom website or e-mail to contact me for any reason or for study suggestions, hints, and guides.

If necessary and I'm not conducting class, you can still stop in anytime. My door is always open.

Helping Students Be More Successful by Checking for Understanding

- Verifies that learning is taking place
- Provides opportunity for student feedback
- Improves instruction

Strategy Examples for Checking Understanding

- Formative: ongoing/reviewing/observing (i.e.)
    - Random Q and A during discussions/tutoring
    - Study guides (not worksheet), puzzles, concept maps, summary charts, reading/vocabulary
    - Labs/activities/projects
    - Quizzes / lab practicals
    - Review games (add trivia to make it more exciting)
- Summative: effectiveness/competency/cumulative (i.e.)
    - Properly created chapter/semester exams

Note: See nonconclusive list to help check for understanding. (See page 58)

Strive for next-day service. (Feedback should be within 24 to 48 hours.)

Addendum

Concept Map

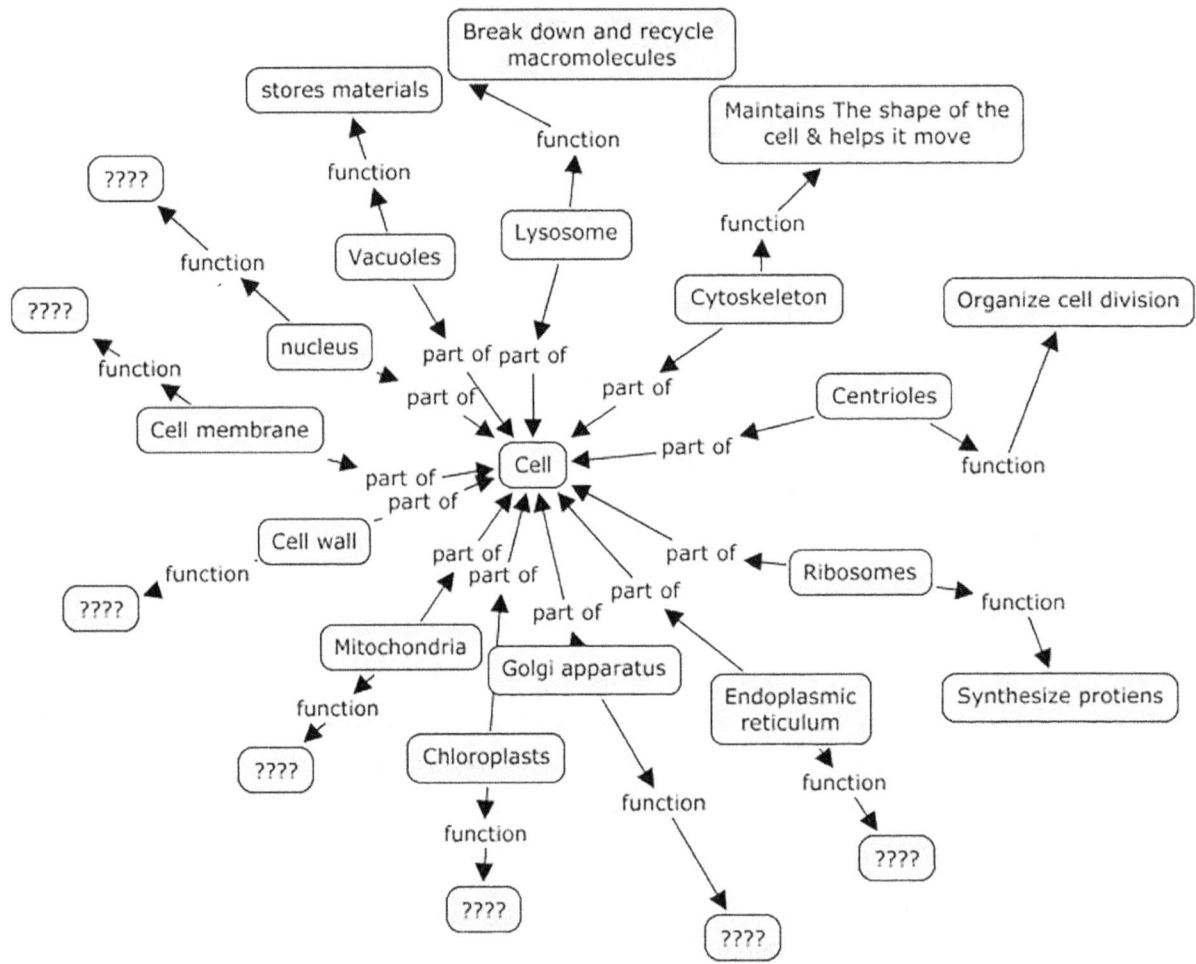

SCIENCE STRATEGIES TO INCREASE STUDENT LEARNING AND MOTIVATION IN
BIOLOGY AND LIFE SCIENCE GRADES 7 THROUGH 12

Addendum

First and Last Name:

/75

Cell Structure is Related to Cell Function Summary Chart
1 Point Per Box Below

Organelle or Structure	Describe the Structure(s) *(Shape/Size/Chemical Make-up/etc...NOT Analogies)*	Describe the Main Function(s) *(Few Words; Yet Descriptive...NOT Analogies)*	Found In Animal, Plant, or Both Cells
Nucleus (Nucleolus/DNA)			
Cell Membrane *(not an organelle; but a structure)*			
Cytoplasm *(not an organelle; but a solution)*			
Cell Wall *(not an organelle; but a structure)*			
Floating Ribosome(s)			
Attached Ribosome(s)			
Rough Endoplasmic Reticulum			
Smooth Endoplasmic Reticulum			
Golgi Body (Apparatus)			
Mitochondrion			
Leucoplast			
Chromoplast			

DAVID BUTLER

Chloroplast			
Lysosome			
Vacuole			
Cytoskeleton *(not an organelle; but a structure)*			
Spindle Fiber(s) *(not an organelle; but a structure)*			
Centriole(s) *(not an organelle; but a structure)*			
Centrosome *(not an organelle; but an "area")*			
Cilium (Cilia) *(not an organelle; but a structure)*			
Flagellum (Flagella) *(not an organelle; but a structure)*			

Organelles with 1 Membrane   Organelles with 2-3 Membranes   Organelles with No Membranes

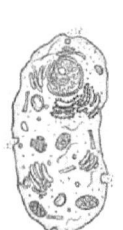

List Here	*List Here*	*List Here*
(6 Total from the List Above)	*(4 Total from the List Above; 3 of the 4 Similar)*	*(2 Total from the List Above but Similar Names)*

(12 Points Possible Above)

SCIENCE STRATEGIES TO INCREASE STUDENT LEARNING AND MOTIVATION IN BIOLOGY AND LIFE SCIENCE GRADES 7 THROUGH 12

Addendum

Topic-Reading Assignment and Vocabulary Evaluation List "Histology"

Reading

1. Read, understand, and complete the following reading assignment by visiting masteringanadp.com.
   Note: Be sure that you have your online text registered and password and username ready. Second, that your device is configured correctly.
   Read chapter 6, "Osseous Tissue and Bone Structure," and complete online assignment.
   Read chapter 10, "Muscle Tissue," and complete online assignment.
   Read chapter 12, "Neural Tissue," and complete online assignment.
   Read chapter 5, "The Integumentary System," and complete online assignment.
   For the assignments, select Assignments from your online text menu in which two main categories may be listed for chapters assigned.
   - Animations and tutorials are optional; however, they help and are recommended to complement lectures, class activities, and all exams. If you do not see the assignments, then you did not register the class ID code; thus, you must do so when logging on or ask in class.
   - Reading quizzes are required and for a grade. The questions are few; however, the point value is high in order to stress the importance of this assignment. The exam is meant for open reading and assignments. No class time will be given.
2. Maintain the honor policy by taking the exam yourself and with no help from others in any way. If concerns arise, consult the teacher right away.
   Note: If the quiz is not completed by the due date, it is your responsibility to notify the teacher when the quiz will be done. A 10% deduction, an alternative assignment, and/or an incomplete grade (zero) may be given if not done by the due date.

Vocabulary

1. Number, define, and write each of the following terms neatly on a separate sheet of paper in complete sentences in accordance to the topics being discussed. Typing, copying, pasting, or sharing is not allowed. Names of people are primarily defined for what they did or how they contributed to the topic.
2. Maintain the honor policy by doing the vocabulary yourself and with no help from others in any way. If concerns arise, consult the teacher right away.
   Note: If the vocabulary is not completed by the due date, a 10% deduction, an alternative assignment, and/or an incomplete grade (zero) may be given if not done by the due date.

1. connective tissue
2. matrix (tissue related)
3. loose connective tissue
4. dense connective tissue
5. blood
6. plasma
7. bone
8. periosteum
9. compact bone
10. spongy bone
11. red marrow
12. yellow marrow
13. osteocyte
14. lacunae
15. collagen
16. Haversian canal(s)
17. osteoporosis
18. ligament
19. cartilage
20. synovial fluid
21. immovable joint (i.e., suture)
22. ball-and-socket joint
23. gliding joint
24. hinge joint
25. ellipsoid joint
26. pivot joint
27. saddle joint
28. skeletal muscle
29. fascicle
30. fiber (muscle related)
31. sarcolemma
32. myofibril
33. Z-band
34. sarcomere
35. myofilament
36. myosin
37. actin
38. cardiac muscle
39. intercalated disc(s)
40. gap junction
41. myocardium
42. smooth muscle
43. origin (of muscle)
44. insertion (of muscle)
45. flexion (of muscle)
46. extension (of muscle)
47. hypertrophy
48. rigor mortis
49. neuron
50. sensory neuron
51. motor neuron
52. association neuron
53. cell body
54. dendrite
55. axon
56. myelin sheath
57. multiple sclerosis
58. nodes of Ranvier
59. synapses
60. impulse
61. subcutaneous
62. epidermis
63. dermis
64. dermal papillae
65. follicle
66. sebaceous gland
67. sudoriferous gland
68. eccrine gland
69. apocrine gland
70. decubitus
71. goose bumps
72. blister
73. erythema
74. pallor
75. jaundice
76. black-and-blue mark
77. tanning
78. cutaneous
79. mucosa
80. serosa

Addendum

Nonconclusive List to Help Check for Understanding

1. Summary Poem Activity
   - List ten key words from an assigned text.
   - Do a free verse poem with the words you highlighted.
   - Write a summary of the reading based on the words you highlighted.

2. Invent the Quiz
   - Write ten higher-order text questions related to the content. Pick two and answer them in half a page.

3. The 411
   - Describe the author's objective.

4. Opinion Chart
   - List opinions about the content in one half of a T chart and support your opinions in the right column.

SCIENCE STRATEGIES TO INCREASE STUDENT LEARNING AND MOTIVATION IN BIOLOGY AND LIFE SCIENCE GRADES 7 THROUGH 12

5. So What? Journal
   - Identify the main idea of the lesson. Why is it important?

6. Rate Understanding

7. Clickers (Response System)

8. Teacher Observation Checklist

9. Explaining
   - Explain the main idea using an analogy.

10. Evaluate
    - What is the author's main point? What are the arguments for and against this idea?

11. Describe
    - What are the important characteristics or features of the main concept or idea of the reading?

12. Define
    - Pick out an important word or phrase that the author introduces. What does this word or phrase mean?

13. Compare and Contrast
    - Identify the theory or idea the author is advancing. Then identify an opposite theory. What are the similarities and differences between these ideas?

14. Question Stems
    - I believe that because…
    - I am confused by…

15. Mind Map
    - Create a mind map that represents a concept using a diagram-making tool (like Gliffy). Provide your teacher/classmates with the link to your mind map.

16. Intrigue Journal
    - List the five most interesting, controversial, or resonant ideas you found in the readings. Include page numbers and a short rationale (one hundred words) for your selection.

17. Advertisement
    - Create an ad, with visuals and texts, for the newly learned concept.

18. Five Words
    - What five words would you use to describe the term or concept? Explain and justify your choices.

19. Muddy Moment
    - What frustrates and confuses you about the text? Why?

20. Collage
    - Create a collage around the lesson's themes. Explain your choices in one paragraph.

21. Letter
    - Explain in a letter to your best friend about a term or concept.

22. Talk Show Panel
    - Have a cast of experts debate the finer points of an issue.

23. Study Guide
    - What are the main topics, supporting details, important person's contributions, terms, and definitions?

24. Illustration
    - Draw a picture that illustrates a relationship between terms in the text. Explain in one paragraph your visual representation.

25. KWL Chart
    - What do you know, what do you want to know, and what have you learned?

26. Sticky Notes Annotation
    - Use sticky notes to describe key passages that are notable or that you have questions about.

27. 3-2-1
    - List three things you found out, two interesting things, and one question you still have.

28. Outline
    - Represent the organization of by outlining the concept being discussed.

29. Anticipation Guide
    - Establish a purpose for reading and create post-reading reflections and discussions.

SCIENCE STRATEGIES TO INCREASE STUDENT LEARNING AND MOTIVATION IN BIOLOGY AND LIFE SCIENCE GRADES 7 THROUGH 12

30. Simile
    - What we learned today is like…

31. The Minute Paper
    - In one minute, describe the most meaningful thing you've learned.

32. Interview You
    - You're the guest expert on *60 Minutes*. Answer the following:
    1) What are component parts of the topic?
    2) Why does this topic matter?

33. Double-Entry Notebook
    - Create a two-column table. Use the left column to write down five to eight important quotations. Use the right column to record reactions to the quotations.

34. Comic Book
    - Use a comic book creation tool like Bitstrips to represent understanding.

35. Tagxedo
    - What are key words that express the main ideas? Be ready to discuss and explain.

36. Classroom TED Talk

37. Podcast
    - Play the part of an expert and discuss related issues on a podcast using the free Easypodcast.

38. Create a Multimedia Poster

39. Twitter Post
    - Define in under 140 characters.

40. Explain Your Solution
    - Describe how you solved an academic solution step-by-step.

41. Dramatic Interpretation
    - Dramatize a critical scene from a complex narrative.

42. Ballad
    - Summarize a narrative that employs a poem or song structure using short stanzas.

43. Pamphlet
    - Describe the key features of the topic in a visually and textually compelling pamphlet.

44. You've Got Mail
    - Each student writes a question about a topic on the front of an envelope; the answer is included inside. Questions are then "mailed" around the room. Each learner writes his/her answer on a slip of scratch paper and confirms its correctness by reading the "official answer" before he/she places his/her own response in the envelope. After several series of mailings and a class discussion about the subject, the envelopes are deposited in the teacher's letter box.

45. Bio Poem
    - To describe a character or a person, write a poem that includes
    Line 1) first name
    Line 2) three to four adjectives that describe the person
    Line 3) important relationship
    Line 4) two to three things, people, or ideas that the person loved
    Line 5) three feelings the person experienced
    Line 6) three fears the person experienced
    Line 7) accomplishments
    Line 8) two to three things the person wanted to see happen or wanted to experience
    Line 9) his or her residence
    Line 10) last name

46. Sketch
    - Visually represent new knowledge.

47. Top Ten List
    - What are the most important takeaways written with humor?

48. Color Cards
    - Red: Stop, I need help.
    - Green: Keep going. I understand.
    - Yellow: I'm a little confused.

49. Quickwrite
    - Without stopping, write what most confuses you.

50. Conference
    - A short focused discussion between the teacher and student.

51. Exit Slip
    - Have students reflect on lessons learned during class.

SCIENCE STRATEGIES TO INCREASE STUDENT LEARNING AND MOTIVATION IN BIOLOGY AND LIFE SCIENCE GRADES 7 THROUGH 12

Helping Students Be More Successful by Having an Appropriate Atmosphere

- Makes it more conducive to learning
- Sets the stage for engagement
- Promotes efficiency and productivity
- Showcases the subject

Strategy Examples for an Appropriate Atmosphere

- Posters and projects
- Displays and specimens
- Clean and orderly
- Safe yet friendly
- Layout

Helping Students Be More Successful by Having a Lesson Plan Outline

- Provides information about assignments, labs, exams, etc.
- Should be simple and straightforward
- Showcases goals/objectives
- Allows students to manage their week
- Eliminates anxiety/guesswork (especially if absent)
- Can offer parents/guardians an overview about a lesson or important dates

Strategy Examples for Lesson Plan Outlines

- Color code / bold / uppercase abbreviations (key)
- Present to class on Mondays
- Post in class as an e-mail attachment, on the classroom website, and via Remind, etc.
- Include lecture topics, labs, videos, exams, assignments, demonstrations, study sessions, reminders, homework, etc.

Lesson Plan Outline and Information

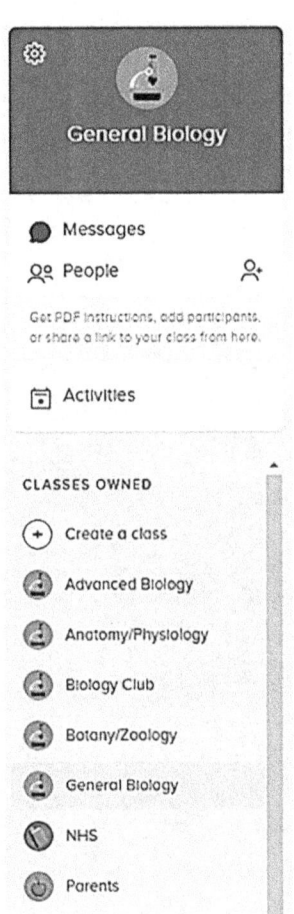

Current Lesson Plan Outline for the Week
Click Link Above / Plan Outline Subject to Change
[.doc or .pdf document]

Learning Strategies and Study Hints

Indiana State Science Standards

Click Links Above

SCIENCE STRATEGIES TO INCREASE STUDENT LEARNING AND MOTIVATION IN BIOLOGY AND LIFE SCIENCE GRADES 7 THROUGH 12

Addendum

Biology Lesson Plan Outline

Week of: State Standards: http://www.doe.in.gov/standards/biology-resources

"Nothing in Biology Makes Sense Except in the Light of Evolution" by Theodosius Dobzhansky

Monday	Tuesday	Wednesday	Thursday	Friday	Next Week
Per 1: Anatomy / Physiology	Per 1: Anatomy / Physiology	Per 1: Anatomy / Physiology	Per 1: Anatomy / Physiology	Per 1: Anatomy / Physiology	Anatomy / Physiology
Lab: Contraction - Q/A DUE by periods end	Lecture: Blood ICA/OCA: Color "Blood"	DUE: Color Lecture: Blood	Lab: Blood - Q/A DUE by periods end	Sub: Video: Heart Attack ~ 30 min ICA/OCA: Review Puzzle	DUE: Puzzle Lecture: Conditions ICA/OCA: SG "Contractions, Blood, and Conditions" QUIZ: Contractions, Blood, and Conditions" DUE: TRAVEL TEST
HW: TRAVEL/WQ	HW: TRAVEL/WQ/Color	HW: TRAVEL/WQ	HW: TRAVEL/WQ	HW: TRAVEL/WQ /Puzzle	DUE: 12/8-9 WebQuest Tissue Trek
Per's 3,6,7: General Biology	Per's 3,6,7: General Biology	Per's 3,6,7: General Biology	Per's 3,6,7: General Biology	Per's 3,6,7: General Biology	General Biology
DUE: Color Lecture: Major Organelles	Lecture: Differences in Cells Handout: Vaccines	Lab: Animal vs Plant Cells	DUE: Lab Q/A Song: Everybody Made of Cells Lecture: Solutions, Membranes, Motion and Transport *Bring Chromebook Friday	Sub: ICA: Color ICA: Summary Chart ICA: SG "Organelles/ Structures" Quiz Monday – no study session	DUE: Color DUE: Summary Chart DUE: SG QUIZ: Organelles/Structures" Lecture: Solutions, Membranes, Motion and Transport OCA: SG's "Cell Diff / Vaccines"; "Sol, Mem, Mot, Trans" Lab: Hyper vs Hypotonic DUE: TRAVEL
HW: TRAVEL	HW: TRAVEL	HW: TRAVEL/Lab Q/A	HW: TRAVEL	HW: TRAVEL/Color /Chart/SG	TEST
Per 4: Botany / Zoology	Per 4: Botany / Zoology	Per 4: Botany / Zoology	Per 4: Botany / Zoology	Per 4: Botany / Zoology	Botany / Zoology
Lecture: Seed Dispersal, Dormancy, Germination Handout – Germination	Lecture: Seedling Anatomy ICA/OCA: Color "M/D Seedlings" ICA/OCA: SG "Pollination → Germination"	DUE: Color DUE: SG QUIZ: Pollination → Germination	Lab: Seed germination - Q/A DUE Monday	Sub: Video: Botany of Desire ~ 120 mins	Video: Cont… DUE: Lab Q/A Mini-Lab: Un-ripe vs Ripe Bananas DUE: TRAVEL TEST
HW: TRAVEL	HW: TRAVEL/Color/SG	HW: TRAVEL	HW: TRAVEL/Lab Q/A	HW: TRAVEL/Lab Q/A	
Per 5: Advanced Biology	Per 5: Advanced Biology	Per 5: Advanced Biology	Per 5: Advanced Biology	Per 5: Advanced Biology	Advanced Biology
QUIZ: Light Independent Reaction TRAVEL Time	Lab: Production of Starch Part II - Q/A DUE by periods end	DUE: TRAVEL Review for TEST – game ?	TEST OCA: TRAVEL	Sub: Video: Journey of Human Life ~45 mins	APA: PPt Handout - Introduction - Headings - Mechanics - Format - Submission - Resources / Guidelines ICA/OCA: SG "APA" Discussion: Concept Maps DUE: TRAVEL QUIZ: APA
HW: TRAVEL	HW: TRAVEL	HW: Study	HW: TRAVEL	HW: TRAVEL	

ICA / ICD = "In Class Assignment/Activity / Demo"
OCA = "Out of Class Assignment/Activity"
HW = "Homework"
SG = "Study Guide"
TRAVEL = Reading/Vocabulary Assignment/Quiz Online
SS = "Study Session"
GD = "Google Drive" Shared Folder

Helping Students Be More Successful with Achievable Goals

- Challenge yet be fair
- Adjust if necessary
- Be realistic
- Obtain with small steps
- Hold students accountable

Strategy Examples to Achievable Goals

- Objective(s) at the beginning/throughout the year
- Checkpoints
- Rubric
- Assess often
- Compliment regularly
- Have students reflect on their work
- Provide students with a curriculum outline

Addendum

Biology Objectives, Materials, Requirements, Information, and Expectations

Objectives

To develop a foundation in biological vocabulary and principles
To develop an awareness about the uniqueness and diversity of life
To develop an appreciation of the interrelationships among living organisms
To develop an understanding of some natural laws and their applications to life
To develop the concept of commonality of structure and function in living organisms

Axioms of Biology (Starting Points of Reason)

Cells
Heredity
Evolution
Energy
Regulation

SCIENCE STRATEGIES TO INCREASE STUDENT LEARNING AND MOTIVATION IN BIOLOGY AND LIFE SCIENCE GRADES 7 THROUGH 12

Addendum

Curriculum Outline and Academic Year Goals/Objectives

Outline for Biology 1 (Lab/General)

(General Biology)

Two semesters, two credits—Grades 9–12: A Core 40 and AHD Course

Class Prerequisite: Pass eighth-grade science.

Required for Graduation (in accordance to Indiana State standards and benchmarks)

Semester No. 1: *Topics Subject to Change / Labs and/or Projects Will Complement Each Area, Subject, and/or Chapter*

Introduction: (*Cellular Chemistry: B.1.1–B.1.3; Evolution B.8.1–B.8.7*)

* Biological Chemistry and Characteristics (*Pyramid of Complexity, Classification, and Defining Life*)
* Lab Methods (*Scientific Method, Tools, Terminology, and Applications*)
* Evolution ([*Spontaneous Generation*], *History* [*Darwinian*], *Principles, Examples, and Evidence*)

Ecology: (*Matter Cycles and Energy Transfer: B.3.1–B.3.5; Interdependence: B.4.1–B.4.7*)

* Ecological Interactions (*Ecosystems, Biotic, Abiotic, and Energy*)
* Energy Flow (*Food Chains and Food Webs*)
* Flow of Matter (*Biogeochemical Cycles*)
* Ecological Limits and Effects (*Capacities, Populations, Stabilities, Fluctuations, Influences, Diversities*)

Cytology: (*Cellular Structure: B.2.1–B.2.6*)

* Histories and Theories (*Early Observations and the Cell Theory*)
* Size, Shape, and Differentiation (*Examples* [*Monera, Protista, and Somatic Cells*]; *Stem Cells*)
* Organelles and Other Components (*Structure and Function*)
* Comparisons of Cell Types (*Prokaryotes and Eukaryotes / Animal Cells vs. Plant Cells*)
* Enzymes (*Reactions*)
* Cellular Fluids and Membrane Functions (*Solutions, States, Diffusions, and Transport*)

Semester No. 2: *Topics Subject to Change / Labs and/or Projects Will Complement Each Area, Subject, and/or Chapter*

Cell Division: (*Cellular Reproduction: B.6.1–B.6.5*)

* Cellular Reproduction (*Asexual and Sexual*)
* Chromosomal Numbers and Structure (*Haploid/Diploid, Histone Complex, Solenoids, Chromosome Anatomy*)
* Cell Cycle (*Interphase [G_1, S, G_2], Cell Division, Cytokinesis*)
* Mitosis (*Prophase, Metaphase, Anaphase, Telophase*)
* Cancer (*Growth Factors, Inhibition, Size, Tumors [Benign/Malignant], Causes, Types, Prevention, Gene Mutations*)
* Meiosis (*Meiosis I and Meiosis II, Spermatogenesis and Oogenesis, Genetic Recombination*)

Genetics: (*Molecular Basis of Heredity: B.5.1–B.5.6; Genetics: B.7.1–B.7.5*)

* DNA "Deoxyribonucleic Acid" (*Structure and Function*)
* Protein Synthesis (*Fundamentals of Transcription and Translation*)
* Terminology (*Gene/Allele, Heredity, Genotype/Phenotype, Dominant/Recessive, Homozygous/Heterozygous*)
* Mendel's Experiments and Principles (*Dominance, Segregation, and Independent Assortment*)
* Punnett Squares (*Monohybrid, Dihybrid, Trihybrid Tables*)
* Probability (*Principle of Chance, Product Rule, Monohybrid, Dihybrid, Trihybrid, and Polyhybrid*)
* Others (*Unknown Alleles, Test Cross, Incomplete Dominance, Codominance, Multiple Alleles, Polygenic, and Transposons*)

Mammalian Anatomy and Physiology Dissection Lab: (*Rattus norvegicus [Norway Rat]*)

* Tissues
* Circulatory System
* Respiratory System
* Digestive System
* Excretory System
* Nervous System
* Sensory System
* Immune System
* and/or Reproductive System

Note: An optional activity for those who do not wish to participate. An alternative assignment will be given.

Visit http://www.doe.in.gov/standards for details on Indiana standards (select correct grade and course).

SCIENCE STRATEGIES TO INCREASE STUDENT LEARNING AND MOTIVATION IN BIOLOGY AND LIFE SCIENCE GRADES 7 THROUGH 12

Curriculum Outline for Biology 2 (Lab/Advanced/Academic)

(Human Anatomy and Physiology)

Two semesters, two credits—Grades 10–12: A Core 40 and AHD Course

Class Prerequisite: Biology 1 and Chemistry 1 (nonconcurrent) with a *B* or higher highly recommended and preferred.

Semester No. 1: *Topics Subject to Change / Labs and/or Projects Will Complement Each Area, Subject, and/or Chapter*

Introduction: (*AP.1.1–AP.1.4, AP.2.1–AP.2.5 [Primarily Incorporated Throughout Curriculum]*)

* Form, Function, and Adaptation (*i.e., Natural Selection or Trade-Off Phenomenon*)
* Tissues, Organs, and Systems (*How Does Structure Correlate with Function?*)
* Body Size Affects Animal Physiology (*Function/Behavior, Surface Area / Volume, Proportions "Allometry"*)
* Electrolyte and Acid-Base Balance (*Salt Content and Fluid Volumes, pH, and Alkalosis/Acidosis*)
* Homeostasis (*Regulation/Feedback*) and Temperature (*Heat Source, Conservation; Ectothermy/Endothermy*)

Histology: (*AP.4.1–AP.4.5, AP.5.1–AP.5.6, AP.6.1–AP.6.10, AP.3.1–AP.3.4*)

* Connective Tissues (*Loose/Dense, Blood, Cartilage, Bone, Muscle, Nerve, and Epithelial*)
* Bone (*Skeleton; Anatomy/Structure/Disease/Joints*)
* Muscle (*Types: Skeletal, Cardiac, Smooth; Anatomy/Structure*)
* Nerve (*Types: Sensory, Motor, Association; Anatomy/Structure/Disease/Impulse*)
* Epithelial (*Layers: Subcutaneous, Epidermis, Dermis; Conditions; Membranes: Cutaneous, Mucosa, Serosa*)

Digestive System: (*AP.12.1–12.5*)

* Biological Molecules and/or Nutrition (*Review and Discussion*)
* Oral Cavity (*Mouth, Glands, and Pharynx*) and Thoracic Cavity (*Esophagus, Peristalsis, Bolus*)
* Abdominal Cavity (*Stomach, Small and Large Intestine, Pancreas, Liver, Rectum, Various Glands / Chemistry*)
* Conditions, Biological Issues, and Implications (*i.e., Ulcers, Anorexia Nervosa, and Bulimia*)

Circulatory System: (*AP.9.1–AP.9.4, AP.10.1–AP.10.6, AP.11.1–AP.11.6*)

* Heart (*Pericardium, Septum, Chambers, Apex, and Pacemaker*)
* Vessels (*Arteries, Arterioles, Veins, Venules, Capillaries, and Lymphatic*)
* Pathway (*Blood through the System* [*Vessels/Chambers*])
* Contraction, Pulse, and Pressure (*Pacemaker, Systolic/Diastolic*)
* Blood (*Plasma, RBC, WBC, Platelets, Antibodies, Antigens, Blood Types, Rh Factor, Counter Current Heating*)
* Conditions, Biological Issues, and Implications (*i.e., Hypertension, Myocardial Infarction, and Leukemia*)

Semester No. 2: *Topics Subject to Change / Labs and/or Projects Will Complement Each Area, Subject, and/or Chapter*

Respiratory System: (*AP.12.1–AP.12.5*)

* Pathway (*Nostrils, Nasopharynx, Larynx, Glottis, Trachea, Bronchi, Lungs, Bronchiole, Alveoli*)
* Gas Exchange (*O_2/CO_2, Hemoglobin, Oxyhemoglobin, HCO_3*)
* Breathing (*Diaphragm, Intercostal Muscles, "Inspiration"/"Expiration"*)
* Conditions, Biological Issues, and Implications (*i.e., Asthma, Pneumonia, Emphysema, Chronic Bronchitis, and Tuberculosis*)

Excretory System: (*AP.14.1–AP.14.6*)

* Excretion (*Homeostasis*)
* Organs (*Liver, Kidney, Ureters, Urinary Bladder, and Urethra*)
* Kidney Structures/Function (*Renal Artery/Vein, Cortex, Medulla, Renal Pelvis, Nephron[s]*)
* Nephron (*Glomerulus, Bowman's capsule, Proximal Tubule, Loop of Henle, Distal Tubule, Collection Duct, etc.*)
* Conditions, Biological Issues, and Implications (*i.e., Kidney Stones and UTI*)

Reproductive System: (*AP.15.1–AP.15.7*)

* Hormones (*i.e., GnRH, FSH, LH, Oxytocin, Prolactin, Testosterone, Estrogen, Estradiol, and Progesterone*)
* Male (*Structure/Function/Anatomy and Sperm Development*)
* Female (*Structure/Function/Anatomy and Ovum Development and Ovulation/Fertilization/Cycle/Gestation*)
* Basic Embryonic, Fetal, Toddler, and Teen Development (*Trimesters/Birth/Puberty*)

Nervous and Sensory Systems: (*AP.2.2, AP.6.1–6.10*) (*May Be Student Assigned Presentation*)

* Brain, Spinal Cord, Nerves (*Electrical Signaling, Action Potential, Synapse, Peripheral System, CNS*)
* Transduction of Sensory Organs (*Hearing, Vision, Taste, Smell, and Touch*)

Immune and/or Endocrine Systems: (*AP.8.1–AP.8.5* [*Primarily Incorporated throughout Curriculum*])

Mammalian Anatomy and Physiology Dissection Lab: (*Felis catus* [cat])

* Anatomical Systems and Physiology (*Used as a Curriculum Review of Mammalian Anatomy/Physiology*)

Visit http://www.doe.in.gov/standards for details on Indiana Standards (select correct grade and course).

Curriculum Outline for Biology 2 (Lab/Advanced/Academic)

(Advanced Biology)

2 semesters, 2 credits—Grades 10–12: A Core 40 and AHD Course

Class Prerequisite: Biology 1 and Chemistry 1 (nonconcurrent) with a *B* or higher highly recommended and preferred.

Semester No. 1: *Topics Subject to Change / Labs and/or Projects Will Complement Each Area, Subject, and/or Chapter*

Biological Molecules

* Terminology (*Molecular Phrases*)
* Hydrocarbons and Alcohols (*Liner, Chain, Ring Components, Hydroxyl Groups*)
* Lipids (*Monomers, Fats, Phospholipids, and Steroids*)
* Carbohydrates (*Monomers, Sugars, Starches, and Cellulose*)
* Proteins (*Functions, Monomers, Primary, Secondary, Tertiary, and Quaternary Types, Enzymes, E_a, Inhibitors*)
* Nucleic Acids (*Monomers and General Structure*)

Cellular Metabolism, Fermentation, and Respiration

* Terminology and Functions (*Enzymatic Hydrolysis, Anabolism/Catabolism, ATP / NAD+*)
* Anaerobic Glycolysis (*Metabolism*)
* Anaerobic Fermentation (*Lactic Acid* [*Lactate Formation*] *and Alcoholic* [*Ethanol Formation*], *Examples*)
* Aerobic Respiration (*Pyruvate Conversion and Krebs Cycle and Electron Transport Chain*)
* Chemiosmotic Theory (*ATP Production*)

Photosynthesis

* Terminology and Functions (*Heterotrophs/Autotrophs, ATP/NADP+, Light-Dependent/Independent Reactions*)
* Plastids (*Chlorophyll and Carotenoids*) and Light Energy (*Nature and Absorption*)
* Photosystems (*Antenna Complex/Reaction Center, Systems I and II, Photophosphorylation, Z-Scheme*)
* Calvin Cycle (*Fixation, Reduction, Regeneration*), *Calvin-Benson Cycle* (C_3) *Pending Discussion*

APA (American Psychological Association) Research Paper

* Guidelines and Instructions on Developing a Paper on a Biological Topic (*due by* year's end)

Semester No. 2: *Topics Subject to Change / Labs and/or Projects Will Complement Each Area, Subject, and/or Chapter*

APA Paper: Briefly *revisited with periodical assignments throughout semester* (*due by semester's end*).

DNA, RNA, and Protein Synthesis

* DNA (*History, Structure, Function*)
* DNA Replication (*Protein and Enzymatic Preparation and Assembling of Complementary Strands*)
* RNA (*Structure, Functions, and Types* [*mRNA, rRNA, tRNA*])
* Transcription "Eukaryotic" (*Initiation, Elongation, Termination, pre-mRNA Editing Phases*)
* Translation "Eukaryotic" (*Codons/Anti-Codons, Ribosome; Initiation, Elongation, Termination Phases*)

Genetic Expression

* Gender Determination of Various Species (*Drosophila melanogaster, Humans, SRY [TDF] Gene*)
* Gender-Linked Traits (*X-Linkage, Barr Bodies, Y-Linkage*)
* Cytogenetic Disorders (*Nondisjunction, Chromosomal Alterations, Gene Mutations*)
* Human Genome (*History and Meaning*)
* Gene Silencing and Influences (*Imprinting, Epigenetics, RNAi, Hardy-Weinberg Principle, Gene Frequencies*)

Organism Diversity/Development: (Student Driven Presentation[s])

* Moneran, Protista, and Fungi (*Characteristics of Bacteria, Protozoan, and Fungi*)
* Plants (*Relationships between Plant Groups, Algae, Lower Plants, and Higher Plants*)
* Animals (*Defining Five (5) Major Features of Animals and Compare Features among Animal Groups*)

Evolution: (Presentation[s] and Discussions)

* Charles Darwin (*Galápagos Islands, Origin of Species, Evolutionary Evidence*)
* Historical Principles (*Mutation and Natural Selection*)
* Kinship Similarities (*Convergences, Fossils, Anatomy and Physiology, Embryology, Genetics, Transitions*)

Curriculum Outline for Biology 2 (Lab/Advanced/Academic)

(Botany and Zoology)

Two semesters, two credits—Grades 10–12: A Core 40 and AHD Course

Class Prerequisite: Biology 1 and Chemistry 1 (nonconcurrent) with a B or higher highly recommended and preferred

Semester No. 1 (Botany): *Topics Subject to Change / Labs and/or Projects Will Complement Each Area, Subject, and/or Chapter*

Plant Diversity

* Introduction to Plants (*Definition, Life Cycle, Survival, Early Plants*)
* Bryophytes (*Groups, Life Cycle, Mosses*)
* Seedless Vascular Plants (*Evolution of Vascular Tissue, Club Mosses, Horsetails, Ferns*)

* Seed Plants (*Gymnosperms and Angiosperms, Cones/Flowers, Seeds, Fruit, Introduction to Monocots and Dicots*)

Roots, Stems, and Leaves

* Plant Tissues (*Epidermal, Vascular [Xylem/Phloem], and Ground [Parenchyma, Collenchyma, Sclerenchyma]*)
* Structures, Types, Adaptations (*Anatomy and Physiology of Mainly Roots, Stems, and Leaves*)
* Water and Nutrient Transport (*Root Pressure, Capillary Action, and Cohesion-Tension Theory*)
* Hormones and Plant Growth (*Auxins, Cytokinins, Gibberellins, Ethylene*)

Reproduction and Development of Seed Plants

* Gymnosperms (*Cones, Pollination, Fertilization, and Development*)
* Structure of Flowers (*Sepals, Petals, Stamens, and Carpels*)
* Angiosperms (*Pollination and Fertilization*)
* Seed and Fruit Development (*Dispersal, Dormancy, Germination*)

Semester No. 2 (Zoology): *Topics Subject to Change / Labs and/or Projects Will Complement Each Area, Subject, and/or Chapter*

Introduction to Zoology

* Introduction to Animals (*Definition, Invertebrates/Vertebrates, Survival*)
* Evolution (*Cell Specialization and Levels of Organization, Early Development, Body Symmetry, Cephalization*)

Sponges and Cnidarians

* Form and Function (*Body Plan, Feeding, Respiration, Circulation, Excretion, Response, Reproduction*)
* Ecology of Sponges and Cnidarians (*Temperature, Water Depth, Light, etc.*)
* Groups of Cnidarians (*Jellyfishes, Hydras, Sea Anemones, and Corals*)

Worms and Mollusks

* Flatworms (*Form and Function: Feeding, Respiration, Circulation, Excretion, Response, Reproduction; Types*)
* Roundworms (*Form and Function: Feeding, Respiration, Circulation, Excretion, Response, Reproduction; Types*)

SCIENCE STRATEGIES TO INCREASE STUDENT LEARNING AND MOTIVATION IN BIOLOGY AND LIFE SCIENCE GRADES 7 THROUGH 12

- * Annelids (*Form and Function*: *Feeding, Respiration, Circulation, Excretion, Response, Reproduction*; *Types*)
- * Mollusks (*Form and Function*: *Feeding, Respiration, Circulation, Excretion, Response, Reproduction*; *Types*)
- * Ecology of Worms and Mollusks (*Aerating / Mixing of Soil, Food Provision*)

Arthropods and Echinoderms

- * Arthropods (*Evolution*; *Form and Function*: *Feeding, Respiration, Circulation, Excretion, Response, Reproduction*)
- * Groups of Arthropods (*Crustaceans, Spiders, Mites/Ticks, Scorpions, Insects* [*Form and Function*])
- * Echinoderms (*Form and Function*: *Feeding, Respiration, Circulation, Excretion, Response, Reproduction*)
- * Groups of Echinoderms (*Sea Urchins, Sand Dollars, Brittle Stars, Sea Cucumbers, Sea Stars, Sea Lilies*)
- * Ecology of Arthropods and Echinoderms (*Devastations, Distributions, and Predation*)

Non-Vertebrate Chordates and Vertebrate Chordates

Fishes and Amphibians

- * Non-Vertebrate Chordates (*Definitions, Characteristics, and Groups*)
- * Fishes (*Definition, Evolution, Forms and Functions, Groups* [*Jawless, Sharks, Bony Fish*]; *Ecology*)
- * Amphibians (*Definition, Evolution, Forms and Functions, Groups* [*Salamanders, Frogs/Toads, Caecilians*]; *Ecology*)

Reptiles and Birds

- * Introduction (*Definition and Evolution*)
- * Forms and Functions (*Temperature, Feeding, Respiration, Circulation, Excretion, Response, Reproduction*)
- * Groups of Reptiles (*Lizards, Snakes, Crocodilians, Turtles/Tortoises, Tuataras*)
- * Groups/Types of Birds (*Pelicans, Birds of Prey, Parrots, Cavity-Nesting, Perching, Herons, Flightless Birds*)

Vertebrate Chordates: Mammals (Student-Driven Lesson[s]) (TBA)

Helping Students Be More Successful by Varying Instruction

- Addresses different learning styles
- Prevents stagnation
- Higher interests
- Makes lessons more appealing
- Promotes creativity

Strategy Examples to Vary Instruction

- Technology/multimedia
- Manipulatives
- Hands-on activities
- Incite discussions
- Games for review
- Collaborate with peers
- History/background/stories
- Plan a day in threes or more
- Scaffolding

Typical week of manipulatives

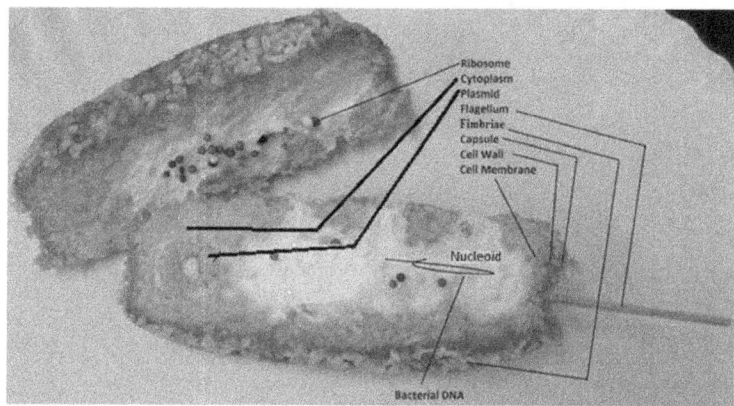

Zinger bacteria model for an introductory lesson on prokaryotes for general biology

Addendum

Scaffolding is used to move students progressively toward stronger understanding and, ultimately, greater independence in the learning process. Teachers provide successive levels of temporary support that help students reach higher levels of comprehension and skill acquisition that they would not be able to achieve without assistance. Like physical scaffolding, the supportive strategies are incrementally removed when they are no longer needed, and the teacher gradually shifts more responsibility over the learning process to the student. One of the main goals of scaffolding is to reduce the negative emotions and self-perceptions.

Example of an end product of *scaffolding* is a student-driven vascular seed plant lab (i.e., various excerpts) after about two to three months on the introduction of plants and plant diversity and a discussion on the use of laboratory equipment and the scientific method for a botany class.

Venus Flytrap

Introduction

How does the Venus flytrap's eating habit affect the plant's metabolism? The Venus flytrap plant is only located in North and South Carolina. The soil in which the plant grows is acidic, and nutrients that other plants thrive off are scarce in the flytrap's soil. The flytrap still uses photosynthesis but has evolved into a carnivorous plant.

Procedure

Part 1: Poisoning

In this part of the lab, two of the four flies will be covered in poison.

1. Take a plastic bowl and spray a thin layer of Dead Sure insecticides. Make sure to wear gloves.
2. Then take the tweezers and carefully dip one fly and leave in the Dead Sure for ten seconds. Repeat that process for the second fly that will be covered in the Dead Sure.
3. Make sure to separate the poisoned flies and the unpoisoned flies by keeping them in different plastic bowls.

Part 2: Feeding

In this part of the lab, one plant will be fed poisoned flies, and the other will be fed normal flies.

1. Take the two flies that were not dipped in the Dead Sure and on one plant, trigger two or three of the sensitive hairs on two of its leaves with the flies.
2. Repeat that process for the two flies that were dipped in Dead Sure and feed them to the second plant.

Part 3: Recording

1. For flytrap 1 and 2 record daily results by writing down the changes the plant has went through until the flies are fully digested by the flytraps.
2. Include numerical observations, as well as descriptive observations.

References

David Andrew. 2012. "In Search of Plants in the Wild." Botany Today 55, (2012): 24–34.
Croasdale, Hannah, Walter E. Loomis, and Carl L. Wilson. 2016. *Botany: Third Edition*. USA: Holt, Rinehard, and Winston Inc.
Jones, Alex. "Occupying Habits in the Southeastern United States of America." Accessed October 20, 2008. http://www.botany.org/Carnivorous_Plants/venus_flytrap.php.
Smith, John. 2015. *Knowing the Venus Flytrap*. Seattle: Macmillan Corporation.
Starr, Mark. 2016. "Carnivorous Mechanics…Are They Real?" Biological Memory and Cognition. Online.

SCIENCE STRATEGIES TO INCREASE STUDENT LEARNING AND MOTIVATION IN BIOLOGY AND LIFE SCIENCE GRADES 7 THROUGH 12

Method

Apparatus

Two (2) Venus flytraps
Four (4) live flies
Light source
Can of Dead Sure insecticide spray
Plastic bowl
Forceps
Gloves
Computer
Resources

Results

After a weeklong research, the findings of this lab were incomplete because both of the plants had gone into dormancy. Just the leaves were dead or dying before the full lab could be performed (see figure 1). The leaves became brown and black in coloration, and both had no responsiveness when the trigger hairs were touched.

Quantitative analysis was very limited due to the fact that the plants died. The only numerical data collected were the number of flies each plant ate: that being zero. Each plant was to have two flies, but when the first flies were placed on the Venus flytrap's leaves, both plants could not eat the prey because of the condition of dormancy. Therefore, none of the plants were poisoned and only showed natural metabolism.

Conclusion

How does the Venus flytrap's eating habit affect the plant's metabolism? When the Venus flytrap eats poisoned insects, the plant's metabolism will be affected and will show signs of death. The hypothesis was refuted. The reason may be because the lab could not be fully performed because of the plant's being in dormancy. Unforeseen problems include not choosing a carnivorous plant that is in the right season. The Venus flytrap should have been kept in the fluorescent light longer. Alternatives to the lab can include making a little greenhouse for the plants so the plants could be in a more humid environment. Would the results have been different if the plants were kept in a more humid environment while out of dormancy?

Increasing Student Motivation in the Biology / Life Science Classroom

Increasing Motivation with Enthusiasm

- It can be contagious
- Provides a level of excitement
- Elevates energy
- Makes reaching objectives/goals more likely
- Helps "sell" the subject

Strategy Examples for Enthusiasm

- Express genuine interest.
- Smile and be cheerful.
- Portray inviting body language.
- Have a stimulating voice.
- Dwell on "Aha!" moments.
- Present with energy and passion.
- Avoid fellow pessimists.

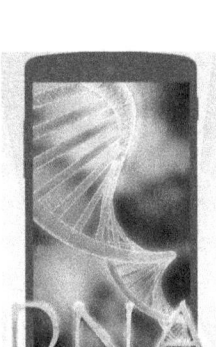

SCIENCE STRATEGIES TO INCREASE STUDENT LEARNING AND MOTIVATION IN BIOLOGY AND LIFE SCIENCE GRADES 7 THROUGH 12

Increasing Motivation with Showmanship

- Provides entertainment
- Demonstrates commitment
- Elevates student interest

Strategy Examples for Showmanship

- Be dedicated to the topic.
- Share your interest.
- Display your accomplishments.
- Your classroom is your stage.
- Role-play.

Reviewing results after a genetics lab

Increasing Motivation with Feedback for Progress

- Provides self-reflection
- Offers opportunities for response
- Checks for understanding
- Demonstrates how to improve

Strategy Examples for Feedback for Progress

- Comment on assignments/labs/exams.
- Timing is everything.
- Avoid being judgmental.

- Give constructive, positive, and specific critiques.
- Mention the "effort made" and "what they accomplished" (*i.e.*, *"You were well prepared for your biology quiz today; your hard work paid off!"*).

Addendum

7. Calculate and give your group's "phenotypic" ratio *(Note: look at your totals; add the two blue totals together first; between the totals of your blue results and the red results, choose which number is the smaller number; divide that smaller number into itself and then into the total sum of the blue results; "round" each number):*

    40:16   (2:1)

8. Is your answer to #7 what you "expected" to get phenotypically, that being 3:1?

    no

9. What are all the possible phenotypes of your offspring from your results?

    red shell, blue shell

10. Do you feel that if you had more time, you would get more precise expected result?

    yes

Conclusion: 10 Points Possible

Clearly write a proper conclusion paragraph and utilize the rules/suggestions previously given in class (ie avoid "pronouns"). Be aware of proper grammar and spelling techniques. Be sure to address the problem question, hypothesis statement, if the hypothesis was proven or disproven, any unforeseen event(s), any improvement(s), and a spring board question (with a "?").

Can a simulation help demonstrate how probability and chance affect the outcome of a genetic cross? If one were to simulate shell crosses, then expected phenotypic and genotypic results between two parents that are heterozygous will be observed. The hypothesis was disproven. An unforeseen event was that more homozygous recessive than heterozygous dominant genotypes appeared. An improvement to this experiment would be to use larger shells. What if students were 5 on twenty minutes instead of ten?

I really like your suggestion to use larger shells :)

well done!

SCIENCE STRATEGIES TO INCREASE STUDENT LEARNING AND MOTIVATION IN BIOLOGY AND LIFE SCIENCE GRADES 7 THROUGH 12

Increasing Motivation with Reasons to Be Successful

- Motivates students to do their best
- Limits classroom management issues
- Can help prevent procrastination
- Able to reach set academic goals
- Increases self-confidence and desire

Strategy Examples for Reasons to Be Successful

- Post a list of biology careers.
- Display and discuss famous scientists.
- Stress importance of mastering past lessons.
- Provide leads to more interesting or exciting topics.
- Discuss current and past advances in biology.

What Can You Do With a Life Science Degree?

Forestry
Animal Behavior
Bacteriology
Food Science / Nutrition
Biotechnology
Marine Biology
Teaching
Zoos
Medicine
Conservation
Entomology (Study of Insects)
Biodiversity Studies
Forensic Science
Science Writer / Journalist
Ichthyology (Study of Fish)
Health Fields
Informatics (Computers in Biology)
Research
Veterinary Medicine
Ecology and Environment

Laboratory Animal Science
Mycology (Study of Fungi)
Botany
Wildlife Conservation / Management
Oceanography
Paleontology
Neuroscience
Agriculture
Immunology
Genetics
Bioengineer
Ornithology (Study of Birds)
Proteomics (Study of Proteins)

Careers
https://www.aibs.org/careers

Note: Careers in biological and life sciences (https://www.aibs.org/careers)

Increasing Motivation with Topic Immersion/Focus

- Engages the student
- Helps promote involvement
- Allows for in-depth learning
- Time for exploration
- Topic-centered lessons can provide better collaboration among teacher and students

Strategy Examples for Topic Immersion/Focus

- Use available resources.
- Ask open-ended questions.
- Have students speak the language of biology.
- Interact with students differently on occasion.
- Make them learn by accident.

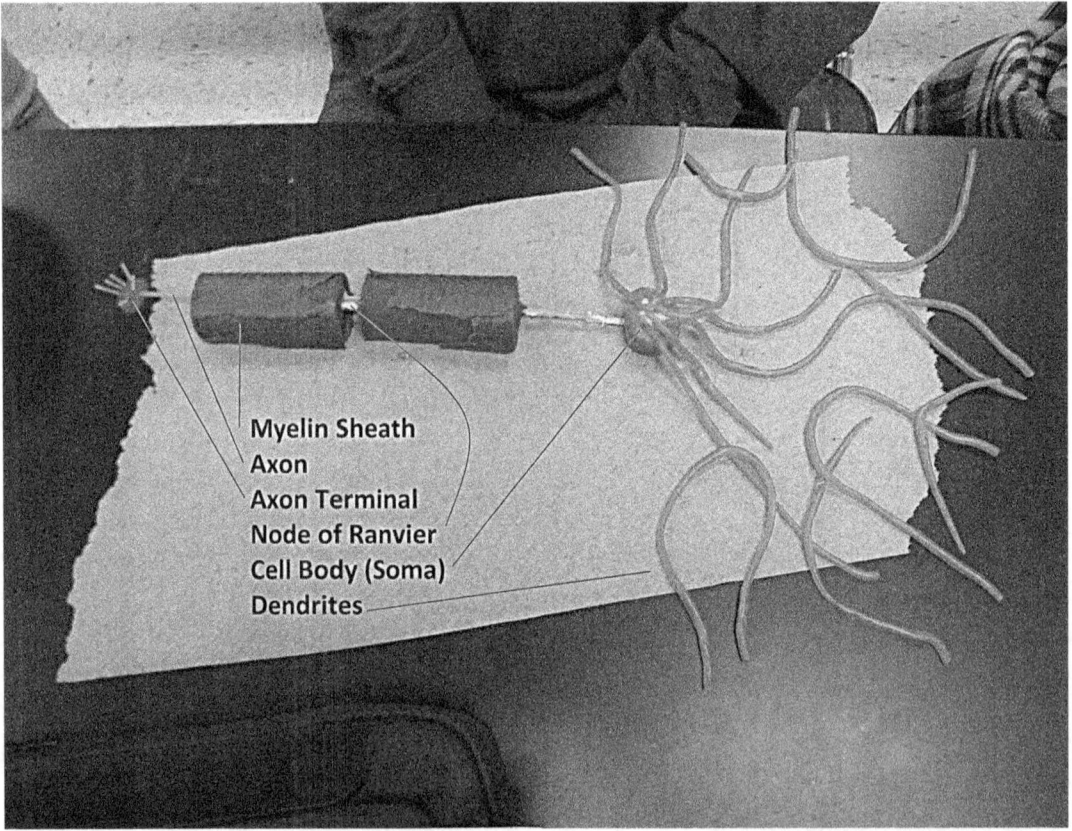

After discussing motor neurons, students had to create a model from HoHos, pasta, candy, and toothpicks and then explain the functions of each part in their own words.

SCIENCE STRATEGIES TO INCREASE STUDENT LEARNING AND MOTIVATION IN BIOLOGY AND LIFE SCIENCE GRADES 7 THROUGH 12

Try to Make Biology / Life Science the Focus Rather than Just Teacher or Student Focused by Adding Variety to Your Lessons

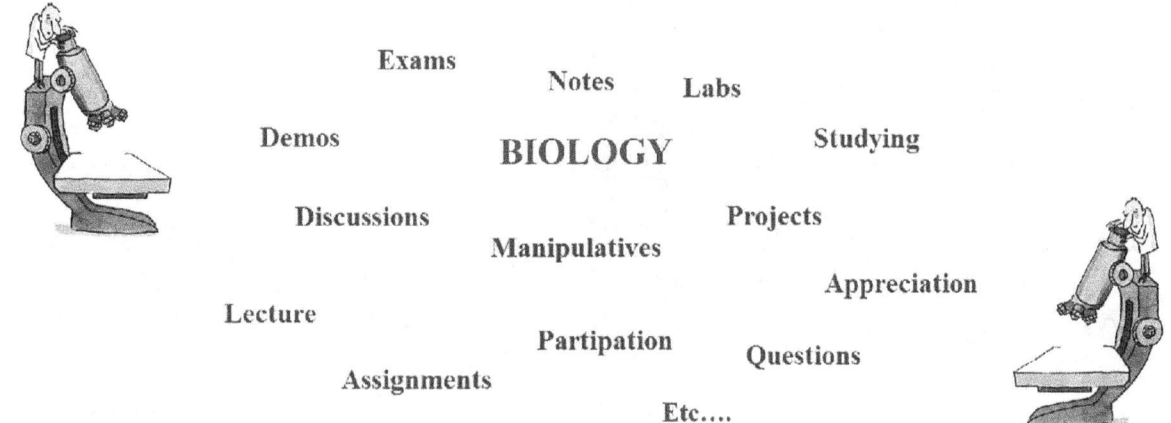

Exams Notes Labs
Demos **BIOLOGY** Studying
Discussions Projects
Manipulatives
Appreciation
Lecture
Partipation Questions
Assignments
Etc….

Increasing Motivation with Student Participation

- Engages students
- Provides feedback
- Permits contribution
- Checks for preparation
- Encourages exchange of ideas

Strategy Examples for Student Participation

- Nonrandom selection
- Random selection
- Small-group discussions
- Assign students to be "experts" in a topic
- Peer-to-peer teaching/tutoring
- "Mole, gopher, or secretary bird"
- Using technology

Advanced biology students "teaching" (not just presenting) about animal and plant diversity to eighth-grade science students

Increasing Motivation with Encouragement and Support

- Promotes self-worth and confidence
- Needs to be positive, optimistic, and hopeful
- You may not know the entire story
- Allows for celebration
- Helps to eliminate frustration

Strategy Examples to Encouragement and Support

- Even a little bit of recognition helps.
- Watch your wording—never be negative.
- Provide guidelines/hints/suggestions.
- Showcase student achievements.
- Focus on reaching realistic objectives/goals.

SCIENCE STRATEGIES TO INCREASE STUDENT LEARNING AND MOTIVATION IN BIOLOGY AND LIFE SCIENCE GRADES 7 THROUGH 12

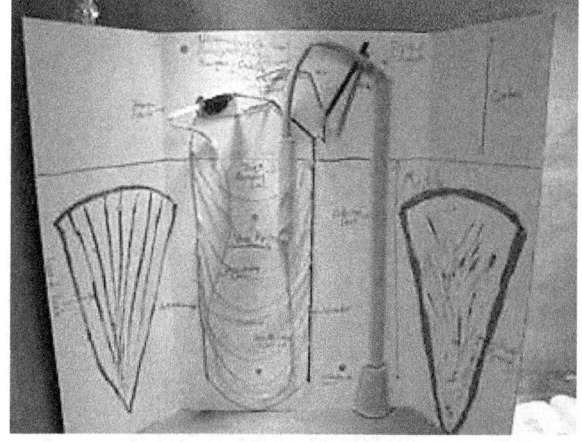

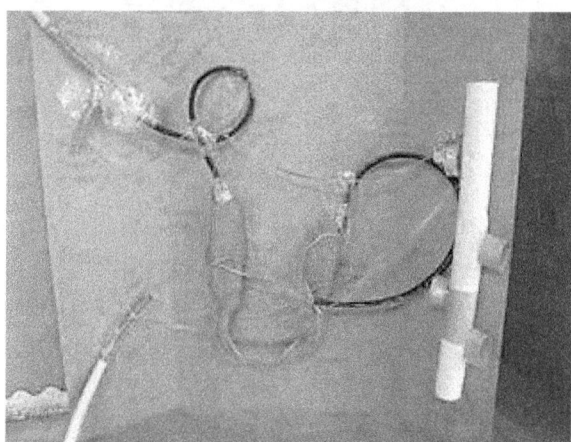

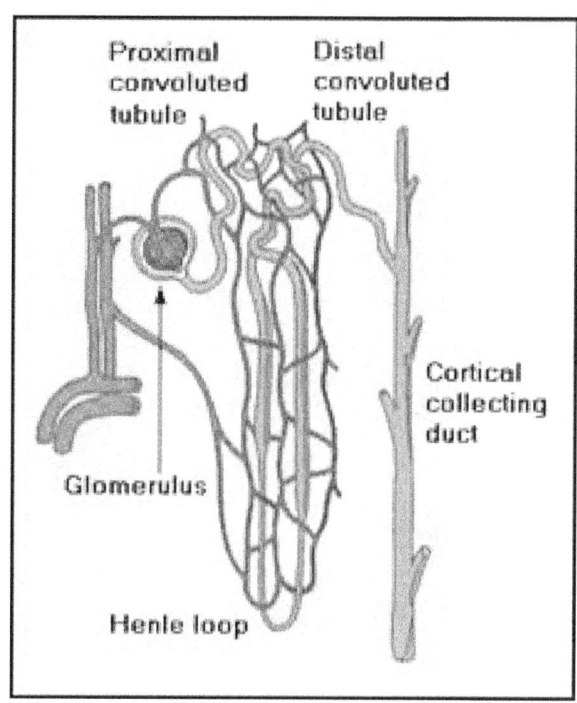

Prior to creating a nephron, students were provided directions, guidance, encouragement, and daily goals even though some models were not perfect...but that's all right.

Increasing Motivation by Highlighting Victories

- Elevates pride
- Displays examples of objectives met
- Promotes student recognition

Strategy Examples to Highlight Victories

- Use past students' above-average examples.
- Showcase not only the best.
- Spotlight partners/groups rather than a student.
- Limelight momentary congratulations.
- Be complementary not derogatory.
- Bulletin boards, websites, social media, etc.

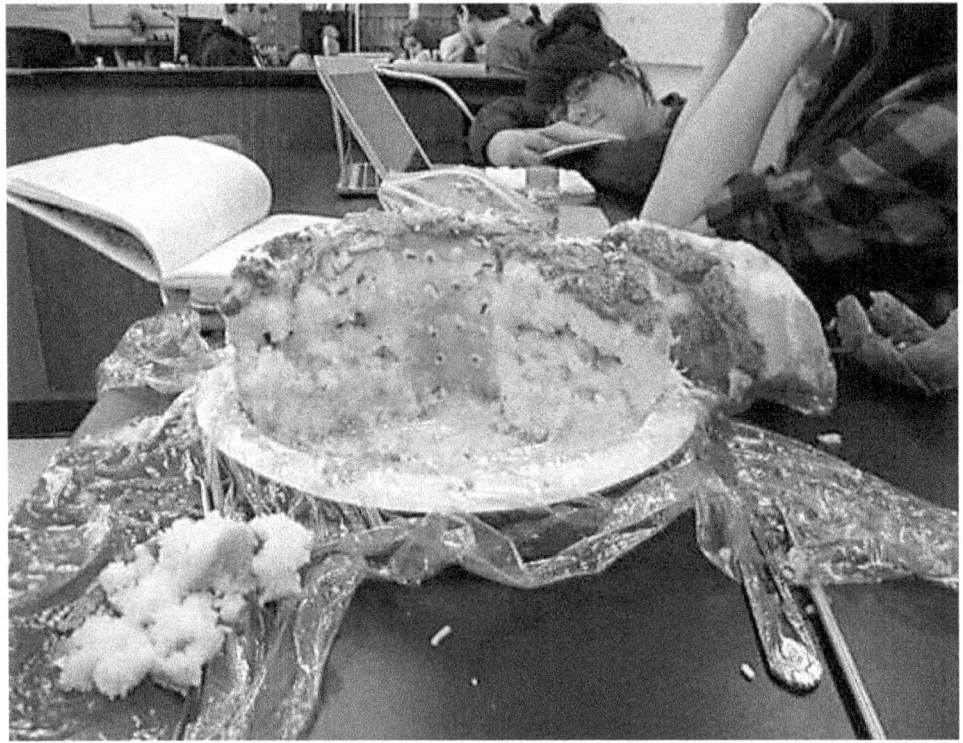

The student in the back with the slight grin on her face suggested a picture be taken of their "sponge cake model" for the classroom web page since her group did well enough to receive a B+ for this review activity on the phylum *Porifera*.

Increasing Motivation by Finding Their Interests

- Makes a connection
- Opens dialogue and discussions
- Reaches outside the classroom

Strategy Examples for Finding Their Interests

- First-day Q and A: Getting to Know You and the Class.
- Talk between classes.
- Acknowledge them at events.
- Find a topic interest between both of you.
- Know their names!

SCIENCE STRATEGIES TO INCREASE STUDENT LEARNING AND MOTIVATION IN BIOLOGY AND LIFE SCIENCE GRADES 7 THROUGH 12

About Me, the Teacher

Favorite color and number: Cobalt blue and 6
Favorite food: Chicken tetrazzini
Favorite TV show: *Amazing Race*
Favorite kind of music: Contemporary jazz / classical / 80s / pop
Favorite movie: *Star Wars* (the 1977 original)
Favorite book: *Greatest Show on Earth*
Favorite animal: Sea dragon
Hobbies: Classic cars, motorcycling, technology, photography, theoretical physics, sci-fi, and fossil hunting
Hometown: Born, September, xxxx, xxxx

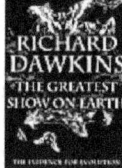

Family: Wife (xxxx), March 8, xxxx - our first date; son (xxxx); daughter (xxxx); dog (xxxx); cat (xxxx)

Education: Graduated from High School in xxxx, Education Major with Biology and Earth/ Space Degrees at BSU, Masters of Education at USI, technology at IUPUI; Other: Adjunct Professor for BSU and ISU

After the students answer "Getting to Know You" questions (*see following*), it's only fair for them to get to know you also.

Addendum

First and Last Name:
Period:

Getting to Know You and the Class

After reading the following short-answer questions, please place your answers in the spaces provided. (*Answers do not need to be complete sentences.*)

- How is your day going so far?
- What is your favorite color *and* number?
- What is your favorite food?
- What is your favorite TV show?
- What is your favorite music and/or group?

- What is your favorite movie?
- What is your favorite book or magazine?
- What is your favorite animal?
- What is your favorite hobby(ies)?
- What city were you born in?
- How many are in your immediate family?
- What would be your dream job?
- What are you most nervous, worried, or concerned about in this class?
- Is there anything else that you would like to share about yourself? (If yes, what else?)

Please state at least one question that you would like answered about the class that will be shared openly. (*Your question will be shared openly in a few moments; however,* no *names will be mentioned aloud.*)

Reading and reflecting on each of the following questions, please place a *Yes* or *No* after each question. (*Add a side comment if you would like.*)

1. Do you believe that it is important to want to come to class nearly every day?
2. Would a variety of different ways of learning help you in class?
3. Would you feel comfortable knowing that if you ever needed help or had a concern, you could talk to or e-mail the teacher?
4. Is it important to you that labs, projects, models, and other class activities be meaningful?
5. Do you feel that technology in the classroom helps you understand concepts being taught?
6. Should a teacher's expectations and goals be fair and reasonable?
7. Should a teacher be fair and reasonable throughout the course with respect to grading?
8. Is the atmosphere (decor) of the classroom appropriate for science?
9. Should the materials and/or topics covered be thorough, adequate, and sufficient?
10. Is it important to you that the exams be professionally written (i.e., wording of questions, variety of questions, detailed questions)?
11. Should the teacher be knowledgeable about the subject matter?
12. Is it important that the teacher be enthusiastic and have an interest in the subject?
13. Do you feel that rules are necessary and that discipline should be handled properly and without bias (unfairness)?
14. Is it important to you that grades be returned and made available promptly and within a reasonable time?
15. Would a classroom website and shared Google Drive folders be valuable and positive features to the course to provide information to students and/or parents?
16. Do you feel that the teacher should treat you with respect and not try to severely embarrass you for the sake of exposing ignorance or unpreparedness?
17. Would you appreciate it if no one would fail the semester as long as their work was completed appropriately and they demonstrate a willingness to try and they maintain a positive attitude?

18. Is it a good idea to have and/or post a weekly lesson plan to help prepare the class and to give direction to exams, labs, projects, and activities?
19. Is it important to maintain good study habits, ask thoughtful questions, stay on task, take good notes, stay organized, do assignments, and ask for help in order to do well in a class?
20. Are you planning to get a *C* or higher for the two semesters of this class this year?

Increasing Motivation by Going Outside the Classroom

- Makes learning more relevant
- Provides a purpose to understand concepts
- Gets students involved with the community
- Can spark interest for an assignment
- Builds student unity and responsibilities

Strategy Examples for Outside the Classroom

- Outside-classroom projects/fieldwork
- Field trips (off *and* on campus)
- Virtual field trips
- Guest speakers
- Participating in community opportunities
- Show-and-tell
- Biology Club

Field Trip to a Botanical Garden

Building DNAs for a Local Light Show

Osteologist Guest Speaker

Incorporating Next-Generation Science Standards to Enhance Biology / Life Science Instruction

"How Can There Be So Many Similarities among Organisms yet So Many Different Plants, Animals, and Microbes?"...NGSS

Addressing: Use Examples, Manipulatives, Analogies, Etc.

- Contemplate time and offer a brief history of life.
- Explain the meaning behind a *theory*.
    - Established facts, testable hypothesis, and confirmed predictions
- Define the benefits of mutations and natural selection.
- Provide evidence for evolution.
- Discuss misconceptions about evolution.
- Compare form and function between organisms.

Strategies to Address, "How Can There Be So Many Similarities among Organisms yet So Many Different Plants, Animals, and Microbes?"...NGSS

Activities

- Fossil Field Trip
- Chinese Whispering (a.k.a. Telephone Game) Simulation
- Phylogenetic Cladograms and Trees Activity
- Darwin's Finches Simulation
- Discuss kinship similarities
- Amazing "Human" Race Activity
- Galápagos Adventure Blog
- Various Comparative Dissections

SCIENCE STRATEGIES TO INCREASE STUDENT LEARNING AND MOTIVATION IN BIOLOGY AND LIFE SCIENCE GRADES 7 THROUGH 12

Fossil field trip in Indiana to find 450–500-million-year-old Paleozoic marine fossils

Addendum

Evolutionary Phylogenetic Cladograms and Trees Activity

Introduction

Phylogenetics is a way to represent the evolutionary relationships within a group of organisms. There are two main pattern types or maps: cladograms and trees.

Is a whale more closely related to a human or a shark? Even though whales and sharks both share certain features of overall body form, by following phylogenetic methods of observation and comparison, we can conclude that whales and humans actually have many more detailed features in common than whales do with sharks. This can be explained by the fact that whales share a more recent common ancestry with humans than they do with sharks. We predict that their closer relationship means that they share more features in common, and the evidence supports this prediction. Similarities between whales and sharks are largely superficial and result from their common aquatic habitat, not from their descent from a common ancestor.

Phylogenetic cladograms are branching diagrams that illustrate patterns of phylogenetic relationships. The pattern of branching itself is the focus of a cladogram; the relative lengths of branches in cladograms have no special significance. Time is included in cladograms only in a relative sense. For example, notice that sharks originated before whales, but we cannot tell from the cladogram how long before (see figure 1). Cladograms are reconstructed by comparing the distribution of characters among species. Characters are inherited attributes of organisms. They may be morphological, genetic, developmental, behavioral, physiological, biochemical, and so forth. Hair color, leg length, or gene sequence, for example, are characters of organisms that vary (see figure 1).

Phylogenetic tree is a branching diagram that illustrates both branching patterns and time. Branch lengths have meaning in the sense that longer branches imply longer periods of time (see figure 2).

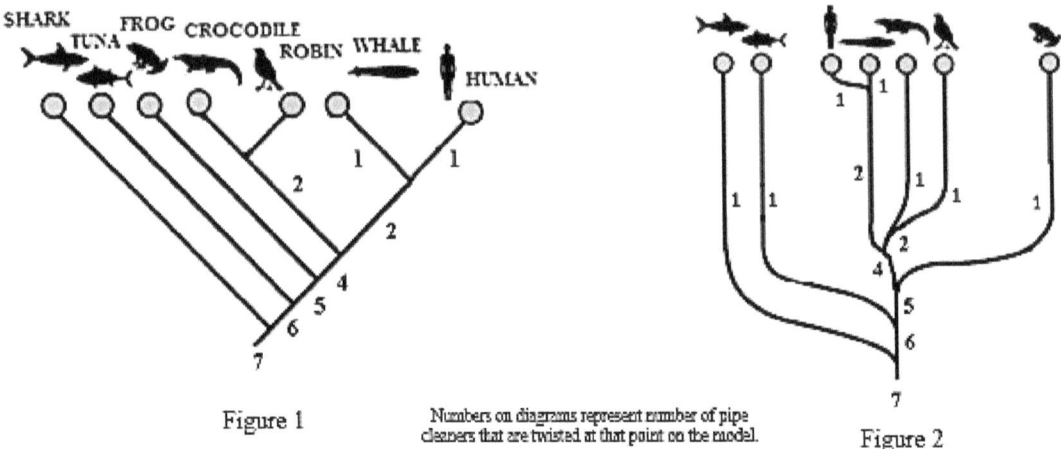

Figure 1

Numbers on diagrams represent number of pipe cleaners that are twisted at that point on the model.

Figure 2

In general, both forms of expression of evolution can be used interchangeably. That being said, cladograms usually have a hypothesis associated with the figure, while trees are usually agreed upon to be more factual.

Procedure

Materials (per student)

Seven (7) different colors of pipe cleaners Colored pencils Scissors
Picture diagrams   Tape   Envelope (optional)

Steps

1. Using colored pencils, color each of the animal pictures neatly, and then using scissors, cut around each animal picture. A circle around the picture is just fine.
2. Decide which type of phylogenetic illustration it is you wish to construct, a cladogram or a tree version. You will use the phylogenetic picture you chose as a guide in making your model.
3. Hold all the pipe cleaners equally distant from one another in a bundle. Again, use the phylogenetic illustration that you chose on the front page as a guide.
4. Tightly twist all the pipe cleaners at the bottom of your bundle to represent the common ancestor that all animals depicted will have.
5. As you move up your bundle, twist the pipe cleaners; however, when you come to a "fork in the road" or a "branch," do not include a pipe cleaner. That pipe cleaner will then repre-

SCIENCE STRATEGIES TO INCREASE STUDENT LEARNING AND MOTIVATION IN BIOLOGY AND LIFE SCIENCE GRADES 7 THROUGH 12

sent a species "'branching" off. Note: Numbers on diagrams represent the number of pipe cleaners that are twisted at that point of the model.

6. Once your model, showing divergences of species, is completed with the pipe cleaners, use tape to attach the animals accordingly at the tips of your phylogenetic map.

Result Questions

Answer the following *or*-questions in accordance to your model, background, and evolution (2 points each).

1. Would you agree that all the organisms share one common ancestor and that they are all related (yes or no)?
2. Does the crocodile have more in common with the *robin, tuna,* or *frog*?
3. Would you agree that *sharks* or *tuna* have been around the longest?
4. Are humans more closely related to the *frog, whale,* or *robin*?
5. Would a *closer* or *more distant* relationship mean that organisms share more common features?
6. According to your model, did the organisms appear *all at once* or *gradually* based on their ancestors?

Table: When creating a phylogenetic model, branches depict differences in characteristics. Place an *X* in the squares that indicate that the following species "mostly" have the characteristic listed (1 point per box). *Research (text/internet) each. (Note: A couple may be tricky.)*

	Hair	Feathers	Lungs	Bipedal	Ectotherm	Symmetry
Human						
Whale	(mostly)					
Crocodile						
Robin					(mostly)	
Tuna						
Frog					(mostly)	
Shark						

Bipedal: walks on two limbs (legs)
Ectotherm: are unable so regulate their own body temperature, cold-blooded
Bilateral Symmetry: only one plane, logical plane, will divide an organism into roughly mirror-image halves

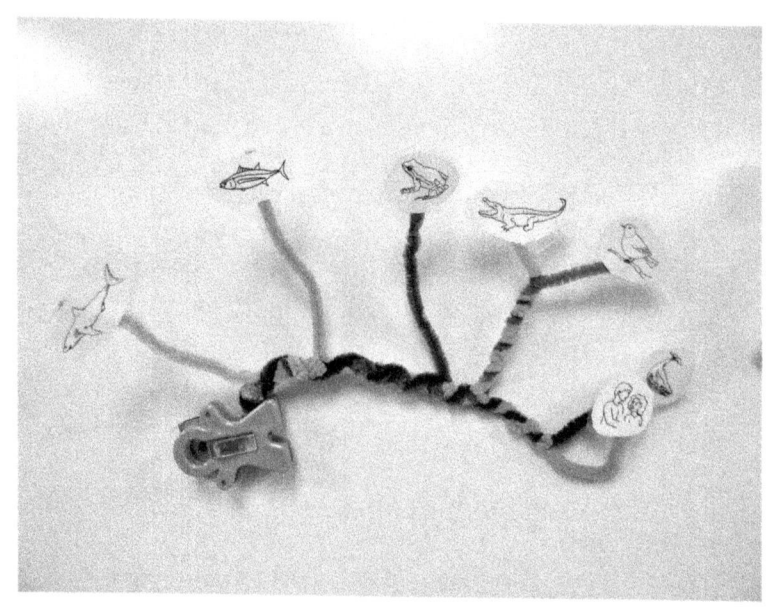

Addendum

Embryology
- i.e. development:
- tails → coccyx
- gill slits → trachea
- arching → spine
- limbs
- symmetry
- yoke sac

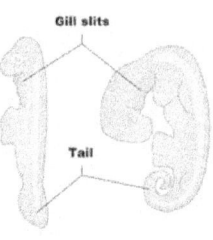
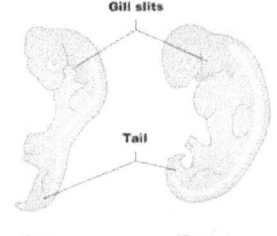
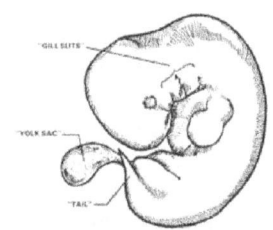

* plus, we all came from a single cell

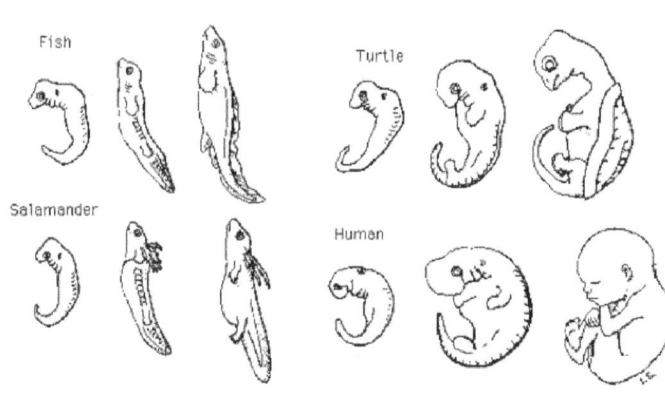
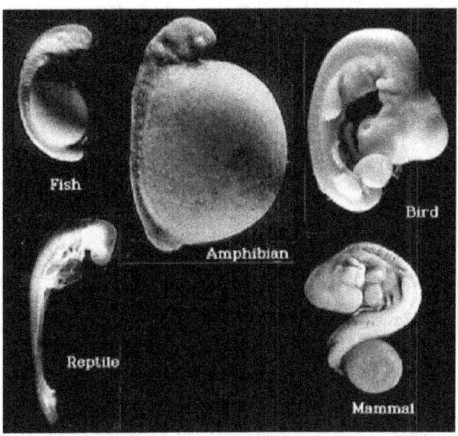

One of many slides used to provide *evidence of kinship similarities* for evolution. Besides embryology, others would include convergence, fossils, genetics, anatomy/physiology, and transitional life-forms.

Addendum

The Amazing "Human" Race

Name:

Background

Human evolution is the lengthy process of change by which people originated from apelike ancestors. Scientific evidence shows that the physical and behavioral traits shared by all people originated from apelike ancestors and evolved over a period of approximately six million years. One of the earliest defining human traits, bipedalism—the ability to walk on two legs—evolved over 4 million years ago. Other important human characteristics—such as a large and complex brain, the ability to make and use tools, and the capacity for language—developed more recently. Many advanced traits—including complex symbolic expression, art, and elaborate cultural diversity—emerged mainly during the past 100,000 years. Humans are primates. Physical and genetic similarities show that the modern human species, *Homo sapiens*, has a very close relationship to another group of primate species, the apes. Humans and the great apes (large apes) of Africa—chimpanzees (including bonobos, or so-called pygmy chimpanzees) and gorillas—share a common ancestor that lived between 8 and 6 million years ago. Humans first evolved in Africa, and much of human evolution occurred on that continent. The fossils of early humans who lived between 6 and 2 million years ago come entirely from Africa.

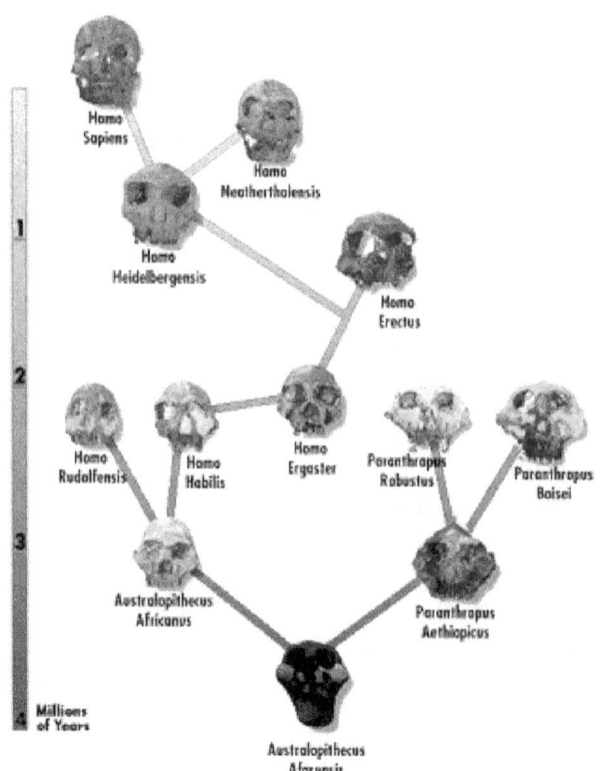

Most scientists currently recognize some 15 to 20 different species of early humans. Scientists do not all agree, however, about how these species are related or which ones simply died out. Many early human species—certainly the majority of them—left no living descendants. Scientists also debate how to identify and classify particular species of early humans and what factors influenced the evolution and extinction of each species.

Early humans first migrated out of Africa into Asia probably between 2 million and 1.8 million years ago. They entered Europe somewhat later, between 1.5 million and 1 million years ago. Species of modern humans populated many parts of the world much later. For instance, people first came to Australia probably within the past 60,000 years and to the Americas within the past 30,000 years or so. The beginnings of agriculture and the rise of the first civilizations occurred within the past 12,000 years. Because of advancements in genetics, only 0.1% of our DNA makes us different from other humans, and only about 2% of our DNA makes our species different from other primates, like chimpanzees. Humans never descended from chimpanzees or other such primates, but instead, we all shared a common ancestor some 7 to 8 million years ago. Our common lineage continues to be made clearer with the help of other biological disciplines besides genetics, such as anatomy and physiology, embryology, and paleontology.

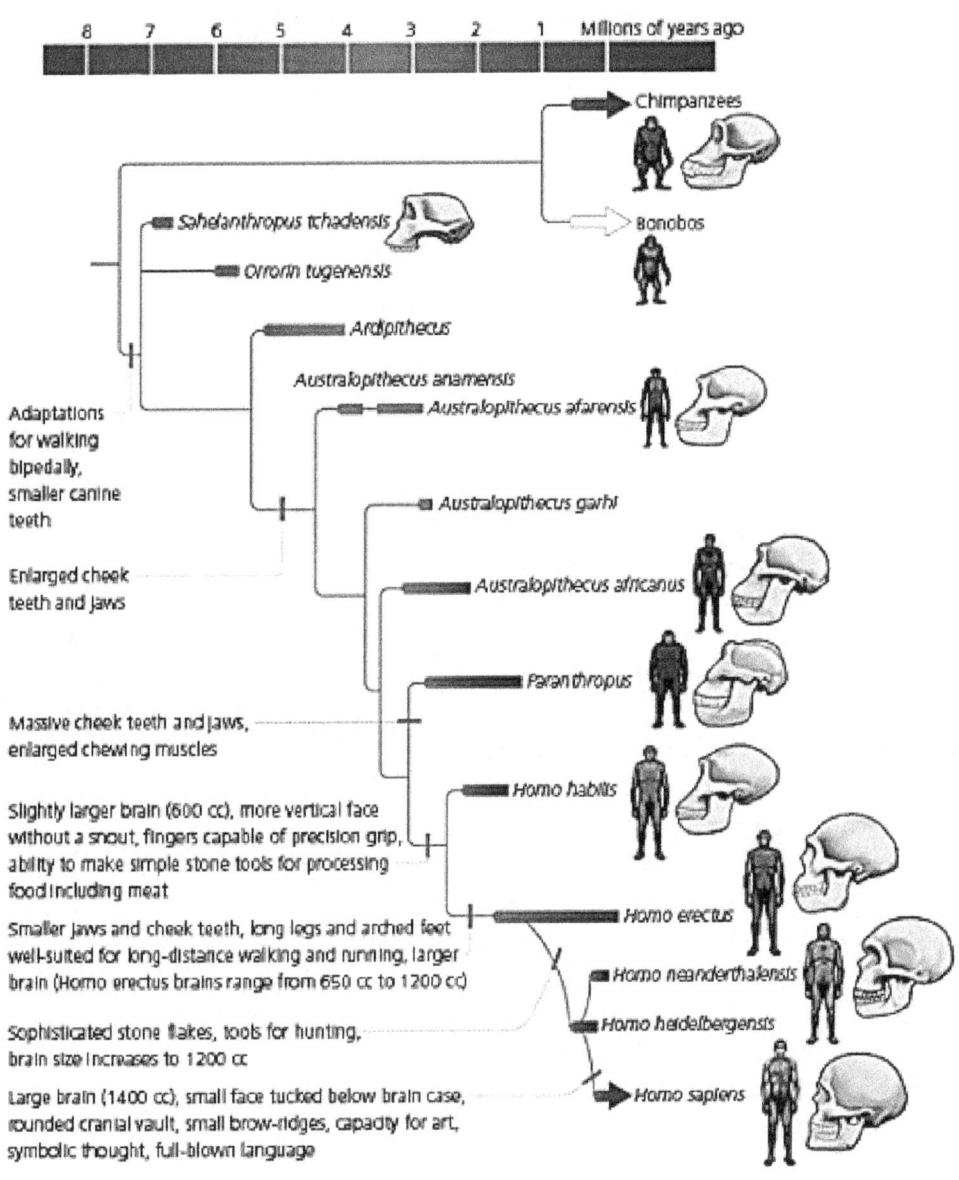

Procedure

Materials

Common ancestor skull replicates:
Sahelanthropus tchadensis (~6–7 million years old)
Australopithecus afarensis (~3.9–3 million years old)
Australopithecus africanus (~2.5 million years old)
Homo habilis (~2.4–1.4 million years old)
Homo heidelbergensis (~700–200 thousand years old)
Homo neanderthalensis (~200 to 29 thousand years old)
Homo sapientes (~300 thousand years old to present day)

Skull cards
Computer/internet resource
Headphones
Metric ruler
Magnifying glass
Caliper cutout (on card stock paper)
Large paper clip (slide-on hinge)
Scissors
Resources

Note: You cannot take the two-minute penalty if you decide not to do a task. All tasks must be attempted.

Steps

Part I: Research "Do Prior to Activity Outside of Class"

1. Develop a basic understanding of human evolution by exploring the following interactive documentary, which will help to explain our origins and our evolutionary journey.
   http://www.becominghuman.org/node/interactive-documentary
   Play, listen, and understand the documentaries for the following titles:
   Prologue, Evidence, Anatomy, Lineage, and Culture.
2. After watching the documentaries, feel free to explore other areas of the website or other evolutionary facts. Note other resources may be and have been provided for you.

Part II: Skull Stations "Done Just Prior to Activity"

1. Create your metric caliper by cutting the diagram out and using a paper clip. Keep this lab, your answer sheet, and the caliper with you throughout the game as "passports." If you leave any one of these items behind, you must retrieve the item before moving on. Note: the metric ruler and magnifying glass stay with the skulls.
2. Cast of hominin skulls have been laid out for you to observe ancestrally human evolution representing a span of millions of years for Part III and the Q/As.

SCIENCE STRATEGIES TO INCREASE STUDENT LEARNING AND MOTIVATION IN BIOLOGY AND LIFE SCIENCE GRADES 7 THROUGH 12

Treat the Models with Respect and Care

Part III: Participating

You are about to embark on an amazing race, the human race. The hominin skulls you will be working with are real molds from real bones found to be thousands and even millions of years old from around the world. The real trip will be to learn about these ancient humanlike and human fossils but also to make you work for the opportunity to learn about your own evolution.

1. You and your partner will "race" around the school to locate a special "flag" (a multicolored paint swatch) that will indicate a staff member (office worker, teacher, administrator, etc.) who will ask you to perform a task. In some cases, it may be a roadblock (in which <u>one of you</u> decides to do the task—cannot switch out) or a detour (in which <u>both of you</u> have to do the task); it's up to the staff member.
2. If you complete the task to the staff's satisfaction, they will give you a signed "skull card"; however, if the "skull card" is <u>not</u> signed, meaning you did not do the task to the staff member's stratification within the allotted time, you will receive a time out <u>penalty of two minutes</u> once you get back to "our" room. Despite how you are doing or what the staff member is saying or asking of you, you must stop after five minutes and collect the "skull card" and get back to the room.
   Note: You cannot do tasks back-to-back; once you finish a task, you must return to the room with the "skull card." If you see another "flag" before or after the task, that's fine, but again, you have to return to the room and complete a station before doing the next task.
3. Once in the room, show the skull card to the instructor. The instructor will return the skull card to you. The skull card will provide a clue to which station you have to complete. At each station, use the skull mold, created caliper, metric ruler, and resources to answer the questions provided for each skull on your answer sheet. Some answers will be objective and collaboratively agreed upon; however, some of your answers may be more subjective and confrontationally argumentative.
   Note: If you and your partner(s) challenge each other's responses, then that is acceptable and justifiable since science is all about finding plausible answers based on facts and testable hypotheses. Questioning prevents ideas and concepts from becoming stagnant and may promote more plausible solutions to problems.
4. Once you complete a station, you're off to find another "flag" (pending unsigned two-minute waiting penalty).
5. In order to win the game and earn 150 points,
   * attempt all tasks,
   * visit all stations,
   * complete all questions and drawing, and
   * perform a final task implemented by the instructor that will summarize the entire experience.

6. Do not run or walk fast. Do not be loud and obnoxious in the hallways. Follow school rules. Areas (countries) to visit may include gym, agriculture building, office, and high school building. If you decide to help each other, that's up to you.

The world of hominids is waiting for you. Good luck and be safe!

Questions: Place all answers on answer sheet only. Note: If skull does not provide the answer, use N/A.

1. What is the *age* of the hominid skull?
2. Do you feel that the angle of the *forehead* (frontal bone) is great, minimal, or something in the middle when looking at the skull in profile?
3. Is a *supraorbital brow ridge* (the ridge of bone above each eye) present?
4. Based on the skull, how would you describe the *eye socket's* shape and size?
5. Is a *sagittal crest* present anywhere along the top of the skull?

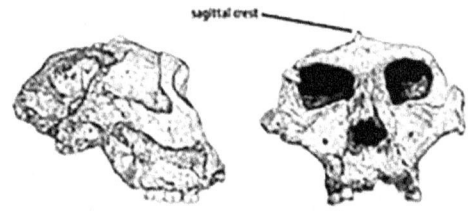

6. What is the shape of the *braincase* (the part of the skull that encloses the brain, the cranium) from front to back when viewed from above?
7. Using the premade calipers and metric ruler (in centimeters), what is the *maximum cranial breadth* diameter (eu-eu)?
8. Using the premade calipers and metric ruler (in centimeters), what is the *bizygomatic* diameter (zy-zy)?

Note:

Euryon (eu): Instrumentally determined ectocranial points on opposite sides of the skull that form the termini of the line of greatest cranial breadth (paired).

Zygion (zy): Instrumentally determined as the most lateral point on the zygomatic arch (paired).

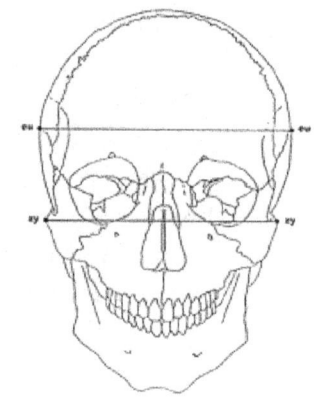

9. Based on the teeth and jaw, would you consider this specimen a *herbivore*, *carnivore*, or an *omnivore*?
10. If present, is the *foramen magnum* (hole at the base of the skull) oriented more downward or more to the rear?

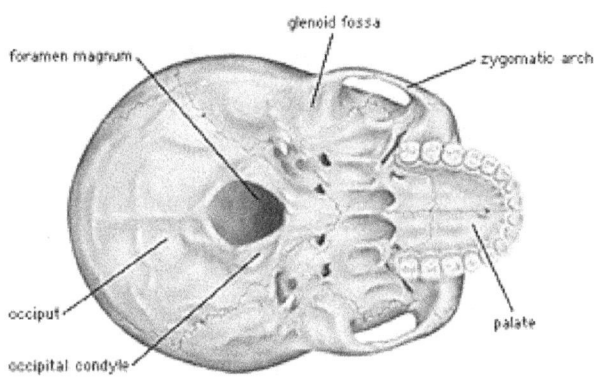

11. Are the *nasal bones* anterior of the nasal opening raised or flat?
12. Do the back *tooth rows* diverge toward the front at an angle? Or are the *tooth rows* more parallel to one another with very little angularity?
13. When viewed from the side, are the *incisors* angled out, angled in, or vertical (straight up and down)?
14. From a scale of *1–7* (1 being the *least* likely), does the skull *resemble* a present-day human model?
15. Based solely on the model (or any information provided), do you think the skull is *female* or *male*?
16. Where was this skull *found* according to information attached to the mold?
17. State at least *one other characteristic* (i.e., speculate on the intelligence level of the specimen) that makes the specimen a predator, or even prey. Or what features make the specimen different or similar to other mammalian primates, such as chimpanzees, gorillas, orangutans, etc.

Evolution of the Skull

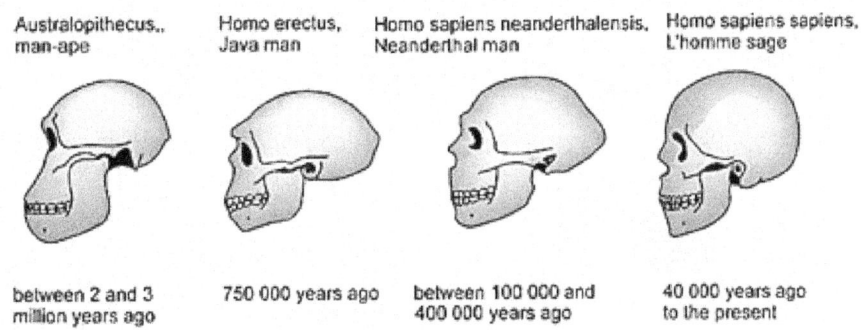

105

Name: _____ Amazing "Human" Race Answer Sheet   /150

Questions and Answer: Use the boxes below to answer the questions from each specimen (1 point each)

Specimen	Age of Hominid	Forehead	Brow Ridge	Eye Socket	Sagittal Crest	Braincase
Sahelanthropus tchadensis						
Australopithecus afarensis						
Australopithecus africanus						
Homo habilis						
Homo heidelbergensis						
Homo neanderthalensis						
Homo sapiens						

Specimen	Cranial Breadth	Bizygomatic	Diet	F. Magnum	Nasal Bones	Tooth Rows
Sahelanthropus tchadensis	mm	mm				
Australopithecus afarensis	mm	mm				
Australopithecus africanus	mm	mm				
Homo habilis	mm	mm				
Homo heidelbergensis	mm	mm				
Homo neanderthalensis	mm	mm				
Homo sapiens	mm	mm				

Specimen	Incisors	I:?	Gender	Location	One Other Characteristic
Sahelanthropus tchadensis					
Australopithecus afarensis					
Australopithecus africanus					
Homo habilis					
Homo heidelbergensis					
Homo neanderthalensis					
Homo sapiens				Europe and Americas	

Drawing: Anytime during the game and using a specimen skull (one) older than 700 ty of your choice, draw an artist's rendition of the frontal or profile pose of the hominid as if they had muscle and skin. (10 points possible). Note if you are found interfering with another team, you will be penalized.

Name of the Specimen Here: _____

SCIENCE STRATEGIES TO INCREASE STUDENT LEARNING AND MOTIVATION IN BIOLOGY AND LIFE SCIENCE GRADES 7 THROUGH 12

"How Do Organisms Obtain and Use Energy They Need to Live and Grow? How Do Organisms Interact with the Living and Nonliving Environment to Obtain Matter and Energy?"...NGSS

Addressing: Use Examples, Manipulatives, Analogies, Etc.

- Define interdependence
- Flow of energy (i.e., food chains/webs)
- Discuss/illustrate biogeochemical cycles
- Give examples of community interactions (i.e., symbiosis)

Strategies to Address "How Do Organisms Obtain and Use Energy They Need to Live and Grow? How Do Organisms Interact with the Living and Nonliving Environment to Obtain Matter and Energy?"...NGSS

Activities

- Food Chain and Food Web Simulations
- Earth-Apple
- TerrAqua Eco-Column

Addendum

Earth-Apple

Earth-Apple

Problem Question: How much of it is soil in which we could grow our food?

1. Cut the apple into four equal wedges longwise (vertically).
   - Three wedges represent the earth's oceans (put aside).
   - One wedge (fourth one) represents the land.
2. Cut the land portion wedge in half longwise (vertically) to make two long slices.
   - One slice is deserts, swamps, Antarctica/the arctic, and mountains.
   - One slice (other half) is where people live and grow food.
3. Carefully cut the where-people-live slice into four smaller thin slices longwise (not as cubes).
   - Three slices are too rocky, wet, hot, developed, or poor soil quality for growing food (put aside).

107

- With the other slice, remove the *outer peel* only. The peel represents the earth's soil on which people depend for food production. This isn't going to change. If anything, it is just going to get smaller.

Addendum

TerrAqua Eco-Column

Introduction

What is an ecosystem? As one of the most encompassing levels of organization in the living world, an ecosystem is defined as a community of organisms, along with their physical environment. The eco-column is designed to model an ecosystem on a small scale. Its components include a terrestrial habitat, with a compost unit, and an aquatic habitat. This system provides opportunities to understand how energy is brought into the living world and transferred through food chains and how the living and nonliving environments are intimately connected through cycles of matter.

Water falls from the atmosphere, flows through our bodies, runs through the soil beneath our feet, collects in puddles and lakes, and then vaporizes back into the atmosphere in a never-ending cycle. As water cycles between land, ocean, and atmosphere form the major link between the terrestrial world (involving anything living on the earth) and the aquatic world (involving anything living on or in the water).

Water drips off rooftops; flows over roads, off your toothbrush, and down the drain; percolates through the soils of fields and forests; and eventually finds its way into rivers, lakes, and oceans. During its journey, water will pick up leaf litter, soil, nutrients, agricultural chemicals, road salts, and gasoline from cars, all of which have profound impacts on life in aquatic systems. Water can also be filtered or purified as it percolates through soil.

The TerrAqua eco-column provides you with a model to explore the link between land and water. The model has three basic components: soil, water, and organisms (plants and/or animals). By varying the treatment of just one of these components, you can explore how one variable can affect the whole system. Once a model of an ecosystem is built a better understand the process of interdependence can be achieved.

SCIENCE STRATEGIES TO INCREASE STUDENT LEARNING AND MOTIVATION IN BIOLOGY AND LIFE SCIENCE GRADES 7 THROUGH 12

Procedure

Materials

Stock Table:
Bottle caps (may be predrilled)
40 lb. (18 kg) bag of topsoil
Cotton clothesline rope (may be precut)
Large spoons (for soil)
Scissors (sharp point)
Cotton balls
Metric ruler
Straws
Masking tape
Cheesecloth
Water
Hole punch
Paper fasteners

Group Responsibilities: Groups of 3–4
1.) Clear and cleaned 2 L soda bottles with cap
2.) Cups (460 mL) of aquatic gravel or very small "cleaned" stones/pebbles
Terrestrial animal(s) (see list)
Terrestrial plant(s) (see list)
Aquatic animal(s) (see list)
Aquatic plant(s) (optional [see list])
Food for living creatures
Method to house any terrestrial animals that may try to escape (i.e., screen / pantyhose / paper wall / plastic dome (using extra bottle), etc.

Steps

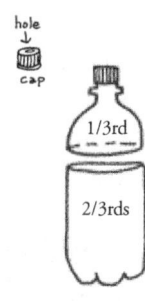

1. Gather all necessary materials for this activity.
2. Drill a hole in the bottle cap while still on the bottle (may have been done already).
3. Remove any labels from the bottle and cut bottle around two-thirds *from the bottom*.
4. Cut around 25 cm length of rope/wick (may have been done already). Thread rope through the bottle top, invert the top, and set the top into the base. The bottom of the wick should reach the bottom of the reservoir, and thread loosely through the cap.
5. Remove the top with the cap/wick and fill the reservoir with enough gravel to just cover the bottom and then add enough water (optional: clean pond water) until the reservoir is about one-third full (or just enough to touch the cap).

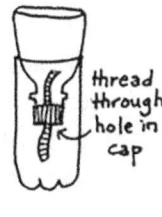

6. Return the top with the cap/wick to its inverted position again. Place a cotton ball down and around the wick.
7. Add a little bit of gravel *and then* add enough topsoil to cover the wick/rope and fill the top chamber almost all the way. To be effective, the wick should run up into the gravel and soil mix and not be plastered along the side of the bottle. Finally, use a straw between the top and bottom reservoir to provide air to the bottom water reservoir area.

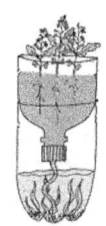

Note: If your top is slipping into the water reservoir, ask for a tool to create 3–4 holes and use paper fasteners to clamp the top to the top of the reservoir.

8. Label your project with your name(s) using a marker and place your eco-column in a designated area and maintain its growth and health (*pick up from the bottom*).
9. After about one to two days, add your organisms. Be sure to add water and take care of any organisms as much as necessary during the allotted time (to be announced).
10. Please clean and return all materials. The result/data sheet will need to be completed after the designated time (to be announced).

Note: Feel free to be creative and/or add a design theme to your column.
Reference: Hunt, Kendall. Bottle Biology. http://bottlebiology.org/index.html

Plant and Animal Suggestion List

(*Common and Scientific Names*)

May have to visit local pet store or a local pond/lake/ground/soil. Remember to supply food for those organisms that need nutrients.

Terrestrial Level: Animals
Note: Have a covering if necessary.
Earthworm, *Annelida*
Sow (pill) bug, *Arthropoda*
Snail, *Mollusca*
Slug, *Mollusca*
Insect, *Arthropoda*
Mite, *Arthropoda*
Beetles, *Arthropoda*
Ants, *Arthropoda*
Crickets, *Arthropoda*

Terrestrial Level: Plants
Mosses, *Bryophyta*
Heartwort, *Ricciocarpus*
Spikemoss, *Selaginella*
Kalanchoe, *Kalanchoe*
Fast plant, *Brassica*
Liverwort, *Marchantia*
Grasses, *Anthophyta*
Pine seedling, *Pinus*
Resurrection fern, *Polypodium*
Seeds (be aware of time allotted)

Aquatic Level: Animals
Betta fish (recommended), *Betta splendens*
Goldfish, *Carassius auratus*
Wheel snail, *Gyraulus*
Pond snail, *Amnicola*
Ghost shrimps, *Ostrocoda*
Water fleas, *Cladocera*, *Daphnia*
Copepods, *Copepoda*, *Cyclops*
Flatworms, *Dugesia*
Mosquitofish, *Gambusia* (<5 cm)
Pond water with micro-invertebrates / protozoa

Aquatic Level: Plants (Optional)
Aquarium bulbs (i.e., Walmart [Sea-Life Inc.])
Dwarf hairgrass, *Eleocharis*
Hornwort, *Ceratophyllum*
Star Grass, *Herteranthera*
Pigmy chain sword, *Echinodorus*
Egeria, *Egeria*
Milfoil, *Myriophyllum*
Duckweed, *Lemna*
Wolffiella, *Wolffiella*
Salvinia water fern, *Salvinia*
Riccia, *Riccia*
Grass leaf, *Sagittaria*
Pond water with algae

SCIENCE STRATEGIES TO INCREASE STUDENT LEARNING AND MOTIVATION IN BIOLOGY AND LIFE SCIENCE GRADES 7 THROUGH 12

Names in Group:

TerrAqua Eco-Column

Brainstorming, Materials, and Responsibility Sheet

In the beginning of this assignment, you will be placed in groups to brainstorm, organize, and plan this project. The handout is to help establish your ideas, materials, and responsibilities. This is a group grade, and everyone needs to put 100% into the project, especially when it comes to accountability. Do not leave your group unprepared. If you agreed to be responsible with materials, for example, you need to make sure you bring those materials, and if absent, have a plan to get them to your group.

1. What materials (nonliving and living) will you need to create your TerrAqua eco-column? *Note: See required materials list in lab to help complete your list.*
2. Confirm who is going to bring which materials for your eco-column by now placing their name or initials next to the materials above as proof of agreement.
3. On the back of this handout, draw and label a sketch of your project based on what you've come up with so far.
4. What idea or theme do you have to make your eco-column distinctive and stand out among everyone else's?

First and Last Name:

TerrAqua Eco-Column Result and Data Sheet

Questions: 2 points each

1. Do you feel that your eco-column represented some form of interdependence? (Yes or No)
2. List one biotic and one abiotic item from your eco-column.
   Note: If something died, the organism is still considered biotic because the plant/animal was once alive.
   Biotic:
   Abiotic:
3. From your eco-column, name a producer (*hint: autotroph*) and a consumer (*hint: heterotroph: carnivore, herbivore, omnivore, or decomposer* [*bacteria, mold, fungus {could be hard to see}*]).
   Producer:
   Consumer:
4. The organisms in your eco-column need energy because energy from light or chemicals (i.e., sugar, fats, and proteins) is constantly being transferred from organism to organism.

Why would your producer be considered an autotroph in order to get energy?

Why would your consumer be considered a heterotroph in order to get energy?

5. Organisms in your eco-column need matter because matter (i.e., H_2O and nutrients [i.e., C, N, and P]) is constantly being recycled throughout the environment by biogeochemical cycles.
Your producer(s) should have absorbed *or* ingested (choose) matter in your eco-column?

Your consumer(s) should have absorbed *or* ingested (choose) matter in your eco-column?

Conclusion Paragraph: 10 points possible (*write clearly*)

* What was the problem question? (*See background paragraph in lab.*) *For example, start off by writing "The problem question was…"* (*ends with a question mark*).
* What was the hypothesis? (*See background paragraph in lab.*) *For example, start off by writing "The hypothesis was that…"*
* Was the hypothesis supported or refuted? If refuted only, tell why? *For example, start off by writing "The hypothesis was…"*
* What unforeseen event(s) happened during the lab? *For example, start off by writing "An unforeseen event was…"*
* What improvement(s) could have been made in the activity? *For example, start off by writing "An improvement for this lab would be…"*
* State a "springboard" question. *Note: The question must relate to the problem question* (*ends with a question mark*).

Figure: 10 points possible (*may use color*)

Every lab should have some form of illustration. Using your project, create a table, chart, or graph (line, pie, bar [no drawings]) in the space provided alongside these directions to help represent your eco-column model. Be sure to label accordingly.

"How Do the Structures of Organisms Enable Life's Functions?"…NGSS

Addressing: Use Examples, Manipulatives, Analogies, etc.

* Outline the hierarchy of life.
* Discuss size, shape, and differentiation.
* Stress the three main structures in eukaryotic cells.
* Relate structure with function (within organelles, between cells, tissues, organs, etc.).
* Highlight organelles and various structures.

- State the differences between eukaryotes and prokaryotes and plant and animal cells.
- Lecture about membranes, various states of cells, and cellular transport.

Strategies to Address, "How Do the Structures of Organisms Enable Life's Functions?"...NGSS

Activities

- Plant vs. Animal Cell Lab
- Bubble Membrane Mini-Lab
- Cell Cake or Haunted/Holiday House Cell
- Shrinky Dink Cells
- Hypotonic and Hypertonic States Using Potatoes Lab
- Phagocytosis/Pinocytosis Peanut Problem

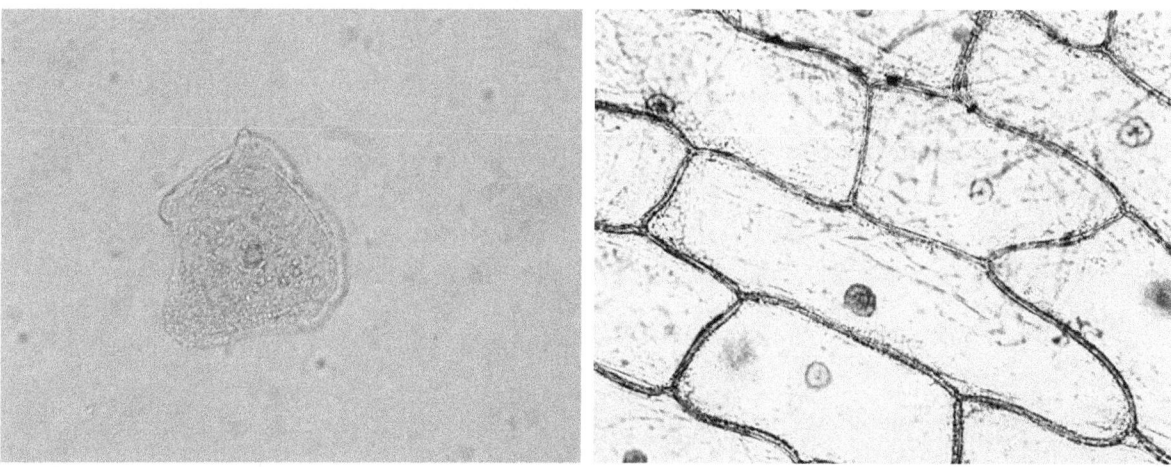

Addendum

Bubble Membrane Mini-Lab

(Simulation of a membrane which shows membrane characteristics.)

Introduction

Bubbles make a great stand-in for cell membranes. They're fluid, flexible and can self-repair. Bubbles and cell membranes are alike because their parts are so similar. If you could zoom down on a cell membrane, you'd see that much of the membrane is a double layer of little molecules called phospholipids. Phospholipids have a love-hate relationship with water. At one end, the "head" is attracted to water, and at the other end, the "tail" is repelled by water. Place phospholipids in water and they quickly form a double layer, with the heads facing out on both sides. A soap molecule has the same split personality. However, with soap membranes, the hydrophobic tails of the phospholipids face *outward* and the heads face *inward*—opposite of what a cell membrane bilayer is in a cell or organelle membrane (see figure).

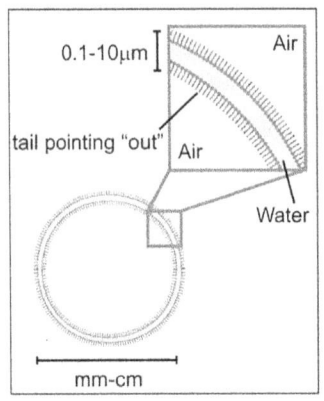
Soap Bubble

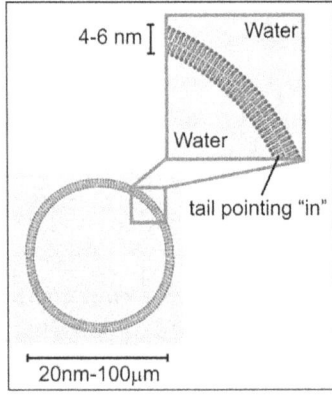
Lipid Vesicle

Materials (Groups of 3–5)
1000mL of bubble solution:
900mL water
100mL dish soap
25mL corn syrup

Tray (i.e., large dissection / cafeteria)
Four (4) bendy straws
Thread (~20 cm [Note: Try a small/light rubber band or light string])
Ruler
Scissors

Steps (Groups of 3–5)

Membrane Bubble Frame

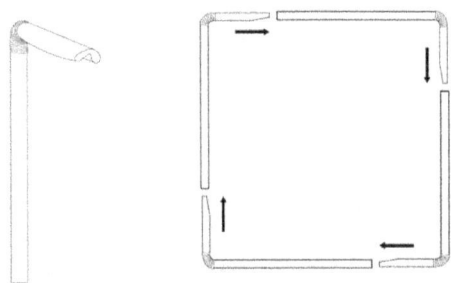

1. Bend 4 straws at elbows.
2. Flatten the shorter ends of the straws and fold flatted surface in the middle.
3. Connect straws together by inserting short ends into long ends to create a square.

Membrane Characteristics: Complete No. 1, No. 2, and No. 3

1. Obtain a tray and place the membrane bubble frame that was just created in it. The frame should be small enough to fit easily.
2. Ask to have just enough soap solution poured in the tray to cover the straw frame.

Characteristic No. 1: Membranes Are Flexible

1. Lift frame out of solution *at an angle* while holding on to the two sides.
2. Tilt the frame back and forth and observe the surface of the film.
3. Notice the swirl of color as the light reflects off the film. Molecules in the cell membrane move about in the same way.
4. Hold the frame by the edges and rotate the sides in opposite directions. Notice the elasticity of the film.
5. Hold the bubble film parallel to the floor and gently move the frame up and down until the surface begins to bounce up and down.

Characteristic No. 2: Membranes Can Self-Repair and Merge

1. Lift frame out of solution *at an angle* so that a thin film spans across the frame.
2. Cover the surface of your hand in bubble solution.
3. Slowly push your hand in and out of the film. The film should allow the hand to pass without breaking because they are covered with the same membranous solution. (This is why structures with similar membranes can merge.)

Characteristic No. 3: Intrinsic Channel Proteins Drift Along Membranes

1. Obtain about 20 cm of heavy thread using scissors. Tie a loop around two fingers and cut off excess thread.
2. Lift frame out of solution *at an angle* so that a thin film spans across the frame.
3. Have someone hold the frame parallel to the tray. Making sure that the thread loop is very soapy, gently lay the loop on the film surface.
4. Use a pencil or pen that *has no soap on it* to break the bubble film that is inside the loop of thread.
5. Insert a pencil or finger into the middle of the thread loop or move the frame around to simulate protein movement.

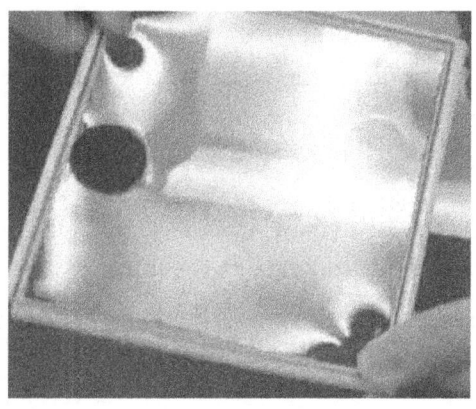

Addendum

Hypotonic and Hypertonic States Using Potatoes Lab

What is the difference between hypertonic and hypotonic conditions? A solution is a homogeneous mixture of two or more substances. One of the substances is called a solvent (what is doing the dissolving [i.e., water is the universal solvent]). The substances dissolved in a solvent are called solutes (i.e., salt or sugar). A solution can exist in a solid, liquid, or gas form depending on the mixed substances and external conditions, such as temperature and pressure. Isotonic solutions are two solutions that have the same concentration of a solute. A hypertonic solution is one of two solutions that has a higher concentration of a solute. A hypotonic solution is one of two solutions that has a lower concentration of a solute. The movement of water from a high concentration to a lower concentration, because the water may be attracted to a solute, is called *osmosis*. When potatoes are used to express hypertonic and hypotonic states, a better understanding of osmosis will be achieved.

Procedure

Materials

Stock:
Distilled water (with 100 mL beakers for distribution)
Granulated sugar (with plastic spoons for distribution)
Group:
Potato
Knife
Spoon (metal)
Plastic plate
Masking tape
References

Steps

1. Write your names and period on the tape and place the tape on your plate.
2. Use a knife to carefully cut the potato into two equal halves as best as possible.
3. Cut a small slice off the bottom of each potato so that both potatoes can sit flat on the plate.
4. Using a metal spoon, scoop out about half of each potato in order to make two (2) potato bowls. Warning: Avoid getting too close to the skin and cutting/poking through the potato.
5. Fill one potato bowl evenly to the top with only distilled water *and* label the potato by placing a piece of tape on the plate and near the water bowl with the letter *W*.

6. Fill the other potato bowl evenly to the top with sugar *and* label the potato by placing a piece of tape on the plate and near the water bowl with the letter *S*. Have one partner take your plate and the other take the filled potato bowls to the designated area *separately*.

"How Are the Characteristics from One Generation Related to the Previous Generation?"...NGSS

Addressing: Use Examples, Manipulatives, Analogies, Etc.

- Lecture on the structure and function of DNA
- Mention the importance of proteins when determining heredity (transcription/ translation)
- Applications and implications of genetic research
- Explain why variation can benefit a species
- Contrast mitosis and meiosis
- Discuss how genes are expressed

Strategies to Address, "How Are the Characteristics from One Generation Related to the Previous Generation?"...NGSS

Activities

- DNA Model Activity
- DNA Extraction Lab
- Protein Synthesis Simulation
- Karyotype and Pedigree Activities
- Phase of Mitosis Lab
- Modeling Meiosis Activity
- Monohybrid and Dihybrid Genetic Crosses Lab

Excited about extracting her DNA

DNA

Addendum

Modeling Monohybrid and Dihybrid Genetic Crosses

Introduction

Can Punnett square simulations help demonstrate how probability and chance affect the outcome of a genetic cross? Genetic crosses come in a variety of different styles. One such cross that involves two parents (P1s) being homozygous or heterozygous for a single trait is called a monohybrid cross. The offspring (F1s) or zygotes of such a cross could also exhibit various genotypes and phenotypes depending on what the parents contribute. The monohybrid Punnett's square is a graphical means to provide a visual understanding of all the possible outcomes between two parent contributions of a single trait (i.e., P1s = Aa x AA); however, a dihybrid Punnett's square would involve two traits (i.e., P1s = AaBB x aaBb).

Probability is defined as the likelihood of an event taking place mathematically. If one were to toss a coin in the air one hundred times, there is a likelihood that seventy tosses or a 70 percent probability may turn out to be heads, and thirty tosses or 30 percent probability may turn out to be tails, even though mathematically, one would expect the coin tosses to be nearly fifty-fifty, or one-is-to-one ratio. Chance is defined as describing the possibility of an event. If one were to toss a coin in the air one hundred times, then one might hypothesize that there is a good chance of getting fifty heads and fifty tails because that is what one would expect. Stating chance is less precise than a mathematical expression, but stating the probability by this means can give one a general idea of the results or the results to come.

When addressing probability, two concepts should be addressed. First, the more data one collects, the more likely one will get what they expect. Second, subsequent (later) events should not be influenced by earlier ones. Nature does not always follow a strict pattern of probability because of environmental changes or genetic imperfections. So gathering a significantly large amount of data while avoiding anything that may influence that data is important in research. Expected genotypic and phenotypic results will be observed by simulating a significant number of genetic crosses between two parents that are heterozygous. Can a simulation help demonstrate how probability and chance affect the outcome of a genetic cross? Genetic crosses come in a variety of different styles. One such cross that involves two parents (P1s) being homozygous or heterozygous for a single trait is called a monohybrid cross. The offspring (F1s) or zygotes of such a cross could also exhibit various genotypes and phenotypes depending on what the parents contribute. The monohybrid Punnett square is a graphical means to provide a visual understanding of all the possible outcomes between two parent contributions of a single trait (i.e., P1s = Aa x AA); however, a dihybrid Punnett square would involve two traits. (i.e., P1s = AaBB x aaBb).

Probability is defined as the likelihood of an event taking place. If one were to toss a coin in the air one hundred times, there is a likelihood that seventy tosses may turn out to be heads and thirty tosses may turn out to be tails even though mathematically, one would expect the coin tosses to be nearly fifty-fifty and that those expected results would only be better supported with a large amount of well-documented data. Nature does not always follow a strict pattern of probability due to environmental or genetic influences. However, if one were to collect and correctly utilize a significantly large amount of data, then what one observes (i.e., naturally or as a simulation) should be statistically close to what one would expect (i.e., found in a Punnett square).

Chance can be defined as the probability that a subsequent event will not be influenced by an earlier event. Many times, it is assumed that if one were to play a game long enough, there is a likelihood that one might win (i.e., a lottery). However, just like the lottery, just because a couple has a child with dimples does not mean that their next child will not have dimples; each event is independent of each other. If one were to simulate several crosses, then expected phenotypic and genotypic results between two parents that are heterozygous will be observed.

Monohybrid Simulation

Procedure

Materials

Two (2) blue-colored macaroni shells
Two (2) red-colored macaroni shells
Two (2) Petri dishes
Plastic bowl
Blindfold / modified sunglasses / closed eyes
Calculator
Pencil
References
Instructor: graphic/spreadsheet program (i.e., Excel)

P1 Snails

Bb x Bb

Steps

You will model the random pairing of alleles by choosing shells from petri dishes. The shells will represent the alleles for having blue shell color (dominant B = *blue* shell) and having red shell color (recessive b = *red* shell). State a proper hypothesis now (see answer sheet).

1. Note table no. 1 on your answer sheet and understand how one is to record the results.
2. Place one (1) blue shell and one (1) red shell in each petri dish. This will simulate each parent being *heterozygous* for having a fictitious blue shell; thus, each parent will be genotypically Bb and phenotypically blue. One dish will represent the testis of a male where sperm are created, and the other dish will represent the ovary of the female where oocytes are created.
3. With their eyes covered, have one partner pick up only one (1) shell from each dish in a similar manner and place them in the plastic bowl simultaneously. The partner's hands will represent male and female gametes (sperm and oocyte) traveling, fertilizing, and eventually

fusing together within the oviducts (or uterus) simulated by the plastic bowl to form a zygote and thus an offspring.

4. After the shells have been chosen, have the other partner accurately record the results on table no. 1. After recording the results (i.e., using tick marks [i.e., /]), have the same partner place the shells back into the dishes. To maintain randomness, move the dishes a little to slightly mix up the simulated alleles. When you and your partner are ready, repeat the steps. Remember, the more you do (the more data), the better the results. When about ten (10) minutes is up, everyone is to total their results at the bottom of each column.
5. Have the recorder record the results on the spreadsheet as directed. Then answer all questions on your answer sheet. A class discussion will proceed after everyone is finished.
6. Clean and return all materials as directed.

Dihybrid Simulation

Procedure

Materials

Two (2) blue-colored macaroni shells
Two (2) red-colored macaroni shells
Two (2) spotted pinto beans
Two (2) white great northern beans
Two (2) petri dishes
Plastic bowl
Blindfold / modified sunglasses / closed eyes
Calculator
Pencil
References
Instructor: graphic/spreadsheet program (i.e., Excel)

P1 Snails

BbSs

BbSs

Steps

You will model the random pairing of alleles by choosing shells and beans from petri dishes. The shells will represent the alleles for having blue shell color (dominant B = *blue* shell) and having red shell color (recessive b = *red* shell). The beans will represent the alleles for being spotted (dominant S = *spotted* body) and being unspotted (recessive s = unspotted or white). State a proper hypothesis now (see answer sheet).

1. Note table no. 2 on your answer sheet and understand how one is to record the results.
2. Place one (1) blue shell and one (1) red shell in each petri dish. Then place one (1) spotted and one (1) unspotted (white) bean in each petri dish. Each dish should have four (4)

different alleles being represented. This will simulate each parent being *heterozygous* for having a fictitious blue shell with spotted bodies; thus, each parent will be genotypically BbSs and phenotypically blue and spotted. One dish will represent the testis of a male where sperm are created, and the other dish will represent the ovary of the female where oocytes are created.

3. With their eyes covered, have one partner pick up only one (1) shell and then one (1) bean from each dish in a similar manner and place them in the plastic bowl simultaneously. The partner's hands will represent male and female gametes (sperm and oocyte) traveling, fertilizing, and eventually fusing together within the oviducts (or uterus) simulated by the plastic bowl to form a zygote and thus an offspring.

4. After the shells and beans have been chosen, have the other partner record the results on table no. 2 accordingly and correctly. After recording the results (i.e., using tick marks [i.e., /]), have the same partner place the shells back into the dishes. To maintain randomness, move the dishes a little to slightly mix up the alleles. When you and your partner are ready, repeat the steps. Remember, the more you do, the better the results. When about fifteen (15) minutes is up, everyone is to total their results at the bottom of each column.

5. Have the recorder record the results on the spreadsheet as directed. Then answer all questions on your answer sheet. A class discussion will proceed after everyone is finished.

6. Clean and return all materials as directed.

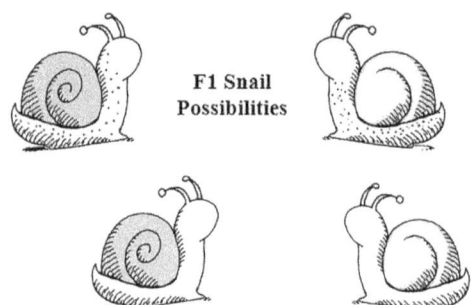

Modeling Monohybrid Genetic Crosses

Table 1: 10 points possible

BB Genotype	Bb Genotype	bb Genotype
Blue Shell Phenotype	Blue Shell Phenotype	Red Shell Phenotype
Total:	Total:	Total:

Questions: 2 points each

1. What fictional trait is being studied or used in this investigation involving snails?
2. What was the genotypic cross between the parents (P1s) according to the given information?
3. What offspring (F1s) genotypes were produced from the P1s?
4. What do your hands, containing the shells (genes), best represent (zygotes or gametes) when being taken from the dishes and placed in the bowl?
5. Calculate and give your group's genotypic ratio. (*Note: Look at your totals. Divide the smallest number into itself and then into each of the rest of your results. Round each number.*)
6. Is your answer to question no. 5 what you expected to get *genotypically*, that being 1:2:1?
7. Calculate and give your group's phenotypic ratio. (*Note: Look at your totals. Add the two blue totals together first. Between the totals of your blue results and the red results, choose which number is the smaller number. Divide that smaller number into itself and then into the total sum of the blue results. Round each number.*)
8. Is your answer to question no. 7 what you expected to get *phenotypically*, that being 3:1?
9. What are all the possible phenotypes of your offspring from your results?
10. Do you feel that if you had more time, you would get more precise expected results?

Conclusion: 10 points possible

Clearly write a proper conclusion paragraph and utilize the rules/suggestions previously given in class. Be aware of proper grammar and spelling techniques. Be sure to address the problem question, hypothesis statement, if the hypothesis was supported or refuted, any unforeseen event(s), any improvement(s), and a springboard question (with a question mark).

Modeling Dihybrid Genetic Crosses

Table 2: 10 points possible

B__S__ Genotype	B__ss Genotype	bbS__Genotype	bbss Genotype
Blue and Spotted	Blue and Unspotted	Red and Spotted	Red and Unspotted
Total:	Total:	Total:	Total:

Questions: 2 points each

1. What fictional traits two (2) are being studied or used in this dihybrid investigation?
2. What were your *phenotypes* as the parents (P1s)?
3. What offspring (F1s) *genotypes* were supposed to be produced? (*Note: Be careful. These are "unknown allele" type genotypes [list on table above].*)
4. What does the bowl containing the shells and beans (genes) best represent (a zygote or gametes) after being placed there by your hands?
5. Calculate and give your group's genotypic ratio. (*Remember, this lab is based on "unknown allele" genotypes. Look at your totals. Divide the smallest number into itself and then into each of the rest of your results. Round each number.*)
6. Is your answer to question no. 5 what you expected to get genotypically, that being 9:3:3:1?
7. Calculate and give your group's phenotypic ratio. (*Note: See title on each column and then look at your totals. Divide the smallest number into itself and then into each of the rest of your results. Round each number.*)
8. Is your answer to question no. 7 what you expected to get phenotypically, that being 9:3:3:1?
9. What are all the possible *phenotypes* of your offspring from your results?
10. Why, in some cases, would a group *not* get a perfect expected set of results?

Graph: 10 points possible

Professionally create a phenotypic/genotypic bar graph based on *your* dihybrid table (*optional color*). Note: You may have to use a straightedge (i.e., ruler) to guide your graph bars.

Number of F1's

Number of F1's	Blue and Spotted B_S_	Blue and Non-Spotted B_ss	Red and Spotted bbS_	Red and Non-Spotted bbss
30+				
29				
28				
27				
26				
25				
24				
23				
22				
21				
20				
19				
18				
17				
16				
15				
14				
13				
12				
11				
10				
9				
8				
7				
6				
5				
4				
3				
2				
1				
0				

Differentiating Instruction More Efficiently in the Biology / Life Science Classroom

Major Learning Style Strategies

- Visual
    - Bold, italicize, underline, and color-code key terms/concepts
    - Manipulatives, models, videos, animations, hand gestures, etc.
    - Outline one's directions, information, and lecture
    - Illustrations on every handout, lab, and presentation

Addendum

Sample PowerPoint slides from lecture discussing the function of ATP synthases during aerobic respiration (ETC)

4. Oxidative Phosphorylation:

* Named this because the production of ATP is powered by the transfer of e-'s from di-nt's to oxygen

ATP Synthase = enzyme that "rotates" when H+'s go through

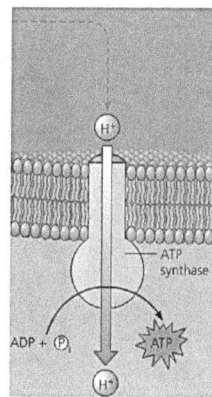

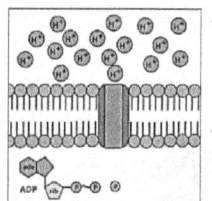

- the rotation transfers energy to ADP + Pi → ATP

- H+'s attach to protein filaments (actin) → causes the rotation; thus, the gradient produced earlier, powers the enzyme → ATP

~ 26-28 ATP's can be made from the ETC

ATP Synthase is like…

Water (ie H+'s) → Dam (ATP Synthase) → Energy (ATP)

- Auditory
- Vocally engage, discuss, and question during lessons
- Read aloud (i.e., study guides, labs, and directions)
- Create rhymes or use music to cement a concept

- Tactile
- Write notes, color assignments, and hands-on activities/labs/projects

SCIENCE STRATEGIES TO INCREASE STUDENT LEARNING AND MOTIVATION IN BIOLOGY AND LIFE SCIENCE GRADES 7 THROUGH 12

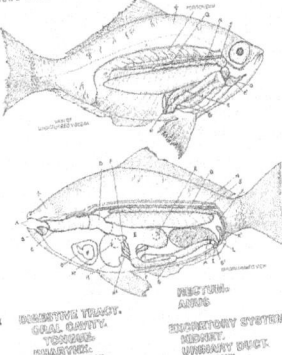

Continual Assessment Strategies

- Walk around the classroom.
- Make sure students understand the material.
- See if students are getting the right answers.
- Exit questions or cards to check for understanding.

Flexible Grouping Strategies

- Allow students to work together in various ways.
- Grouping of students can be determined by:
  - learning preferences—visual/auditory/tactile, quiet vs. loud environment, intelligence types (i.e., analytical, creative, or practical);
  - interests—personal, social, or career;
  - readiness—prior knowledge, work skills, work habits, experience.
- Change it up frequently.

Element Strategies to Help Determine Learning Preferences, Interests, and Readiness

- Content—knowledge/skills students need (i.e., Age of fossils is evidence of evolution.)
- Process—varied activities students can use (i.e., Lectures, labs, videos, or manipulatives about cells)
- Product—students verify their knowledge (i.e., Creating a model of a nephron)

Differentiating involves making changes to one or more of these elements. There is no prescribed way to differentiate. Changes a teacher makes to an element depend on the needs of their students.

Prep Time / Homework / Extra Credit Strategies

- Preparation for differentiation can take time
- One does not have to differentiate every lesson
- Apply certain degrees or levels of differentiation
- One can "water down" formal differentiation by:
    - Limiting homework and/or point values
    - Restricting the amount or value of extra credit
    - Other ways besides bonuses (i.e., bio-bates)

BIO-bate

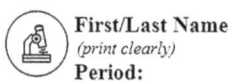 **First/Last Name:**
(print clearly)
Period:

Requirement: *Ask (10) Random People if They Know the Difference Between a Eukaryotic and Prokaryotic Cell. Explain to Them the Main Difference and Have Them Initial the Back Side Acknowledging That They Understood Your Lesson*
Reward: Excused Homework Under 20 Pts
Exp: 12/12

Limit one per student. Void if altered. Assignment necessary. Conditions subject to change. Must be present at time of acquisition. Not redeemable in any other class. Cash value .118 ¢

Making the Best Use of Technology to Enhance Biology / Life Science Instruction

Strategies Using Hardware

- Computers / LCD projectors / DVD players / cell phones
    - Display various websites and files (i.e., Google Drive)
    - Research / review / analyzing and interpreting data
    - Presentations / animations / interactive review sites
    - Show DVDs, Blue-ray, online videos
- Document camera / FlexCam
    - Show manipulatives
    - Ability to work collaboratively
- Smartboard (i.e., various media / interactive software)
- Microscopes, Moticams, and Probeware

Students learning how to use a microscope while following directions and using illustrations and videos on their Chromebooks

Strategies Using "Secure" Software/Sites/Apps

- Presentation software (i.e., PowerPoint/Prezi/infographics)
- Quia / Aurasma / Kahoot! / Socrative / RasMol / BrainPop / Piktochart / Quizizz

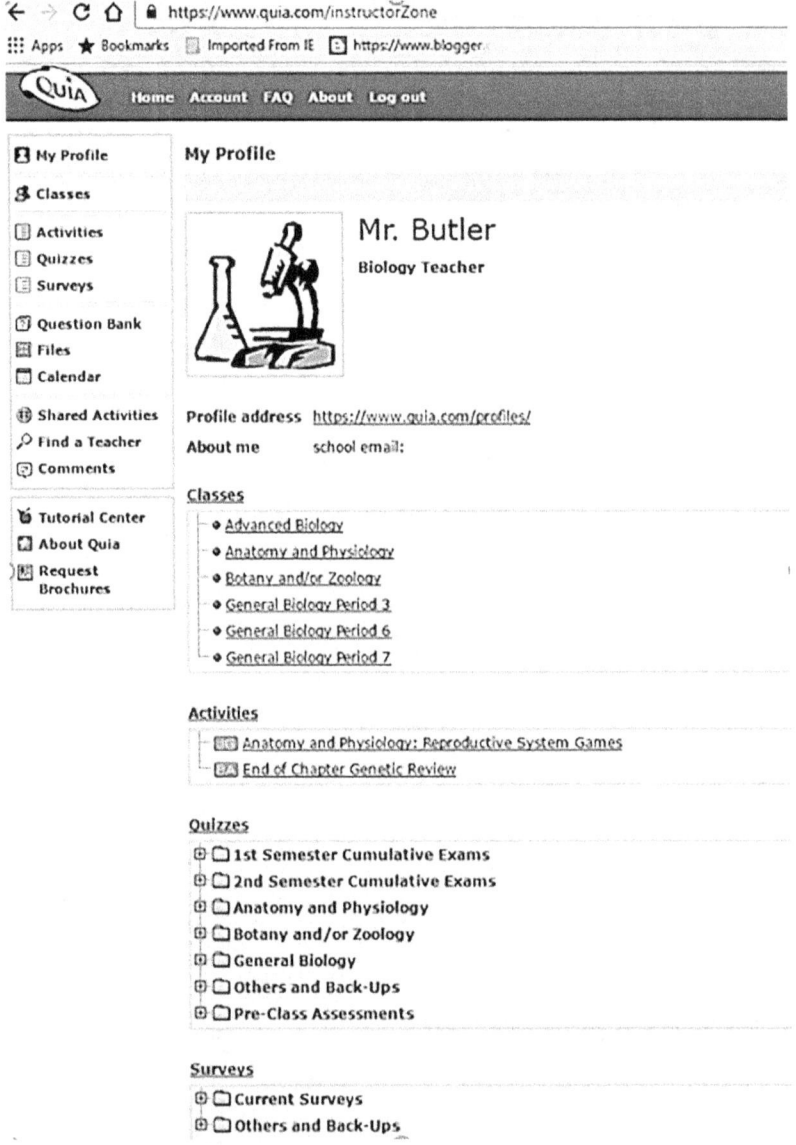

- Storybook maker / online karyotyping / HippoCampus
- VIVED (a.k.a. Cyber-Science 3D)
- *EVO (by Hummingbird Films)*
- Virtuali-Tee *(by Curiscope, UK)*
- CrashCourse / Bozeman / HHMI / MEDLINE / PBS / Khan Academy
- Nucleotide BLAST (basic local alignment search tool)
- Screencast (Google app)
- TeacherTube
- Shape of Life (videos)
- Remind

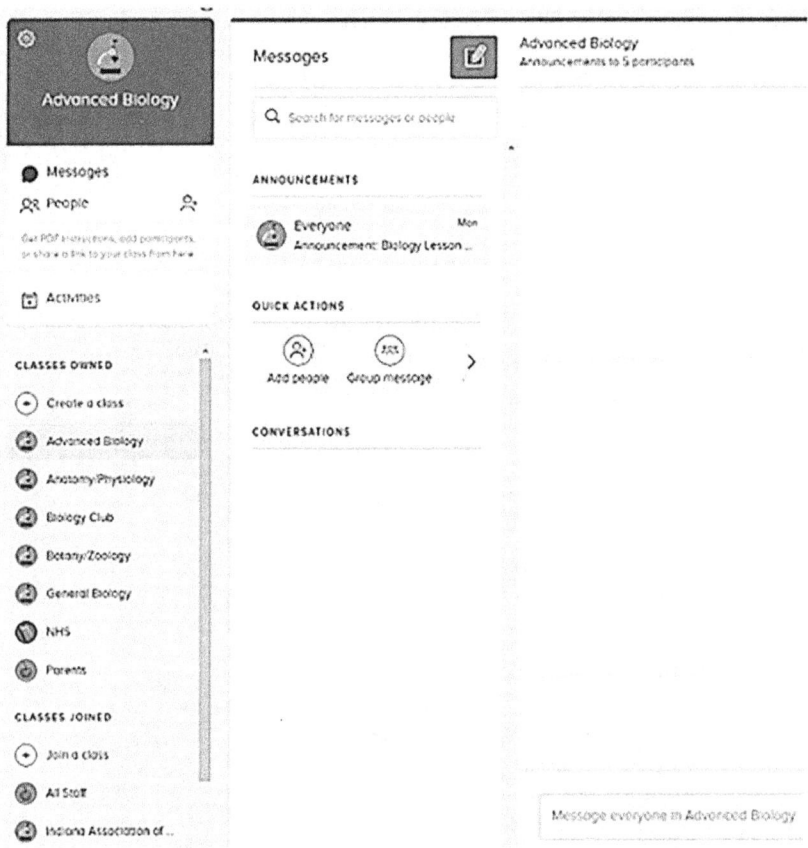

- eText (readings / illustrations / assignments / online exams / simulated labs)
- School e-mail
- Google Classroom / Canvas / Blackboard / Moodle
    - Assignments, labs, articles, illustrations, and tutorials
    - Student work / sharing / monitoring

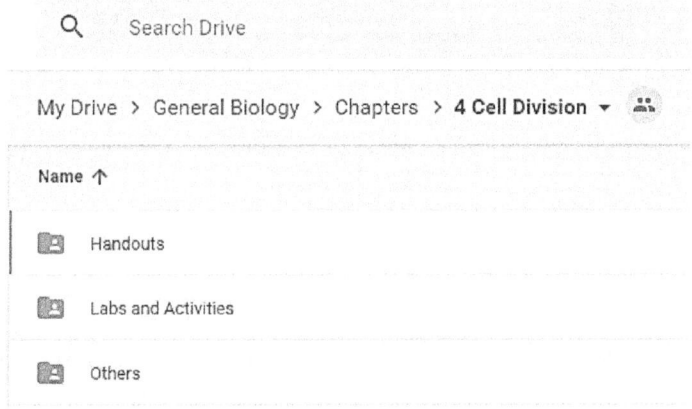

Strategies Using Teacher Created Online Tools

Examples

- Classroom web page
- Biology Club website
- Galapagos Blog

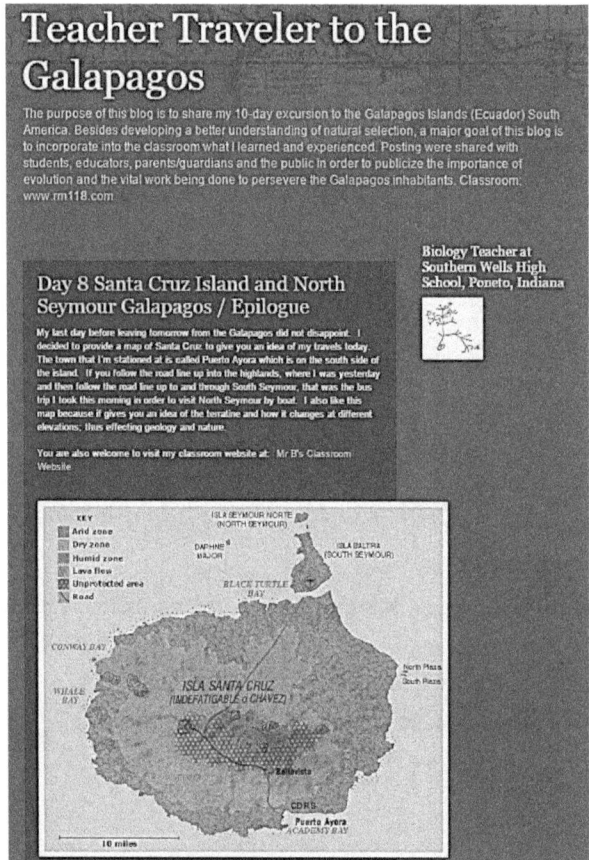

Addendum

Nonconclusive List of Virtually Free Biology / Life Science Sites

Games

- CellCraft: http://www.carolina.com/teacher-resources/Interactive/online-game-cell-structure-cellcr—An interactive online game where students "grow" their own cells. Great way to introduce and discuss cell organelles, evolution, etc. (grades 9–12)

SCIENCE STRATEGIES TO INCREASE STUDENT LEARNING AND MOTIVATION IN BIOLOGY AND LIFE SCIENCE GRADES 7 THROUGH 12

- Nobelprize: aft-biology/tr11062.tr—http://www.nobelprize.org/educational/—A variety of online games about science that discuss a variety of topics. Great use of time when students are finished with classwork (grades 7–12)
- PowerPoint Game Templates: http://bestteacherblog.com/powerpoint-game-templates/—Great and easy website where you can copy and paste PowerPoint game templates right onto your already made PowerPoint. It's more than just Jeopardy.
- Who Wants to Live to a Million Years?: http://www.sciencechannel.com/games-and-interactives/charles-darwin-game/—A great interactive game where students have to choose from a variety of species, survive natural selection, and last a million years. Great way to talk about evolution.
- Arnold Has Lost All His Organs: http://sciencenetlinks.com/interactives/systems.html—An interactive game where students must identify the important parts of the various organ systems. A bit elementary but would be a fun introduction for students and is great for middle school.
- Digestion Story: http://kitses.com/animation/swfs/digestion.swf—A great website to describe what happens to various foods in the digestive system, along with what organs are involved in breaking up simple sugars, fats, and proteins.

Virtual Field Trips and Interactive Websites

- NationalZoo Cam: http://nationalzoo.si.edu/Animals/WebCams/default.cfm?hpout=webcam_link&xtr—Real-life Zoo animal cams
- Virtual Field Trips: http://vft.asu.edu/—Many different virtual field trips
- Learn.Genetics: http://learn.genetics.utah.edu/content/molecules/transcribe/—An interactive activity for students to learn the processes of transcription to translation through a step-by-step interactive lesson.
- BioInteractive: http://www.hhmi.org/biointeractive—A great interactive resource for all students to discover a variety of different science topics from biology, physics, and chemistry. All free!

Videos and Animations

- Nova: http://www.pbs.org/wgbh/nova/
- Open Culture: http://www.openculture.com/science_videos—Video library
- AsapSCIENCE: https://www.youtube.com/user/AsapSCIENCE—Short videos that answer commonly asked science questions
- 21 Smithsonian Scientists: http://www.smithsonianeducation.org/scientist/index.html—A website that has a variety of topics of real-life scientists discovering the world around us.
- SciShow: https://www.youtube.com/user/scishow—A YouTube channel that discusses a variety of topics involving biology, physics, and chemistry subjects.

- TED Talks: https://www.ted.com/talks—A great online resource for teachers to find a variety of topics that the students would be interested in. Also, a very interesting place to look for talks about scientific research, new technology, and more.
- Cell Transport Animations: http://www.scienceandmathwithmrslau.com/2014/11/teaching-about-cell-transport-animations-and-other-resources/—A great blog that has a variety of different cell transport animations such as diffusion, active, and passive transport.

Presentation Outlets

- Emaze: https://www.emaze.com/—A presentation software
- Powtoons: http://www.powtoon.com/—Cartoon presentations
- VoiceThread: http://voicethread.com/—Voice-over video presentations
- Zoho Show: https://www.zoho.com/—An online presentation website

Resource Websites

- Pinterest: https://www.pinterest.com—An online resource for teachers, crafts, and more that can be saved on a personal account
- PhET: http://phet.colorado.edu/—An online simulation resource for a variety of science processes and manipulations that are hard to show in real life
- Earth Exploration: http://serc.carleton.edu/eet/index.html—An online resource for raw data of real-world occurrences (i.e., plate tectonic movement) for earth science, physics, and biology
- Planting Science: http://www.plantingscience.org/—An online resource for teachers to help kids go about making an inquiry project (based on plant science) but can be used for all inquiry projects
- Understanding Science: http://undsci.berkeley.edu/—An online resource for teachers that provide a variety of resources, such as worksheets, PowerPoints, and more about different scientific processes for physics, biology, and chemistry. This website also has age-appropriate subjects from K to 12.
- Understanding Evolution: http://evolution.berkeley.edu/evolibrary/home.php—Similar to Understanding Science, Understanding Evolution is another teacher resource that helps students understand the processes, theories, and developing science of evolution and natural selection.

Others

- Wheel Decide: http://wheeldecide.com/—When deciding what to do with your class, make a wheel to put lab groups together, decide the day of tests/quizzes, and more.

- Field Book Project: http://www.mnh.si.edu/rc/fieldbooks/index.html—This is an online resource where students and teachers can access real-life field journals on subjects such as botany, mammals, birds, reptiles, etc.
- Countdown Timer: http://corkboardconnections.blogspot.com/2012/08/fun-countdown-timers-for-classroom.html—This blog has a variety of different timers any teacher can use for a variety of different things in the classroom, such as quiz-timing, lab time, and more.
- Anatomy Coloring Pages: http://www.momjunction.com/articles/anatomy-coloring-pages_00105691/—A website ran by a mother who has a variety of different anatomy coloring pages for your advanced biology or anatomy/physiology classes. Free to use.
- Introduction to the Class Memes: http://laurarandazzo.com/2014/08/01/meme-me-up-scotty/—Free teacher meme websites for the teacher who likes to involve a bit of humor into the classroom

Site information and URLs are subject to change.

Improving Understanding and Retention of Biology / Life Science Concepts

Improving Understanding and Retention with Music and Dance

- Doug Edmonds or Dr. Art—Singing Scientist
- "Monster Mash"—Ghoulish Glycolysis
- "Bird Is the Word"—Darwin's Finches Lab
- Hand jive dance—Cell Division Review
- Jeopardy theme song—Genetics Review
- "Smooth Criminal"—Fatal Pneumococcus (Griffith's Experiment)
- "You Done Stomped on My Heart"—Circulatory System
- *Amazing Race* theme song—Amazing Human Race
- Contemporary jazz or classical—Labs/Activities

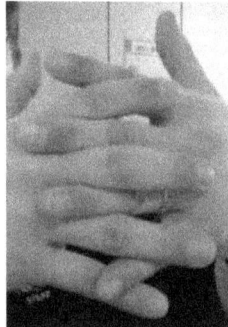

Metaphase: Fingers as Chromosomes

Improving Understanding and Retention with Illustrations

- Presentations—pictures/graphs/charts/animations
- Posters—topic specific during discussions/activities
- Labs/projects—help with directions/concepts
- Study guides—provides assessment/review

Addendum

Fruit flies are very commonly used in genetics because of their size, number, and characteristics. Two characteristics include wing shape and eye size. In fruit flies the allele for normal wings is (W) over the allele for undeveloped wings (w). The allele for normal size eyes (E) is dominant over the allele for small eyes (e).

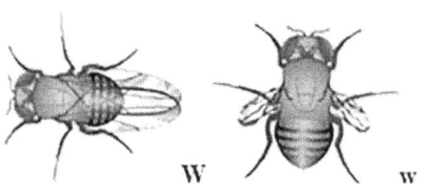
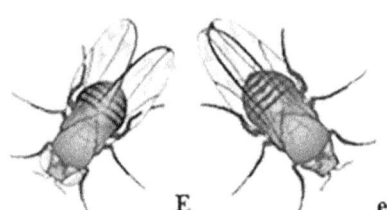

- Board/document camera—supplement lecture
- Picture packet—key visuals used during lecture
- Exams—use for assessment or just a fun visual

SCIENCE STRATEGIES TO INCREASE STUDENT LEARNING AND MOTIVATION IN
BIOLOGY AND LIFE SCIENCE GRADES 7 THROUGH 12

Improving Understanding and Retention with Self-Disclosure

- Stories—Hydrolysis and Dehydration Synthesis
- Personal anecdotes—"Pistil" Packing Mama
- Revelations—Studying Biology by Talking It Out
  - ✓ Can increase focus, interest, and motivation
  - ✓ Provide clarity for a concept
  - ✓ Be careful not to get too much offtrack

Improving Understanding and Retention with Puns and Humor

- We share about 50% of DNA with bananas. Very appealing.
- Mitosis and meiosis? "My toe-iches"; toe is part of the body; thus, involves body cells.
- I find that quiet with regards to ones arm bone.
- rRNA is the translation pirate of protein synthesis.
- Even name blanks:

> "Wood" you mind "leafing" your name "root" here:
>
> **Plant Roots, Stems, and Leaves Exam Answer Sheet**

- ✓ Elevates dopamine → goal-motivator / long-term memory
- ✓ Lowers defenses and brings a class together
- ✓ Keep it topic focused and don't be too "silly"

Improving Understanding and Retention with Mimicry

- Nucleotide—sugar (cupcake); phosphate group (hat); A, T, G, C, U (craft letters)
- Human chromosome—pajamas, belt, socks/gloves, body itself
- Cell membrane—students forming a bilayer with proteins
- Wound—using tissue, glue, and food coloring

Improving Understanding and Retention with Mnemonics

- Bad dog, don't "P" on the MAT
- *K*ing *P*hilip *C*alled *O*nly *F*or *G*ood *S*oup (taxonomy)
- <u>Pu</u>rina Cat/Dog Chow (purine ; food + water ; double-ringed)
- *E*very *M*uscular *T*an *D*ude *S*eems *M*acho (energy, movement, transport, defense, structure, metabolism)
- a MEAGER WORM, C (characteristics of life)
- Hippo (hypotonic vs. hypertonic)
- Eu Have a Brain, Just Like a Eukaryote (having a nucleus)
- Only "AT" spells a word (base-pairing)

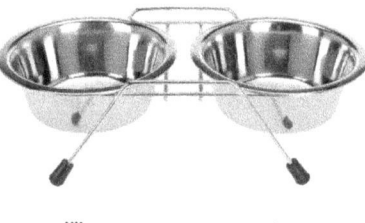

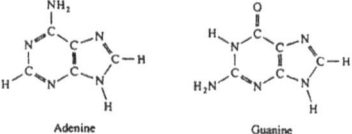

Improving Understanding and Retention with Familiarities/Analogies

- ATP Synthase—Dam (gradient) holding back water (H+s) to generate (enzyme) electricity (energy *ATP*)
- Cell—Nucleus (Wiffle ball), cytoplasm (broth), membrane (bubble), ribosomes (factories), etc.
- Nucleic Acids—DNA (ladder), ATP (rechargeable battery), NADP+ (thief)
- Heart—Two pumps
- Tissue—Bubble wrap, etc.

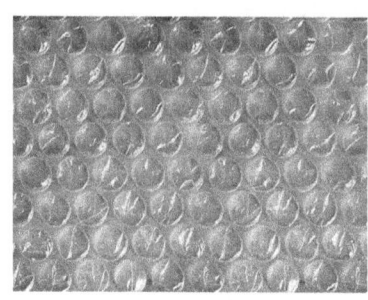

SCIENCE STRATEGIES TO INCREASE STUDENT LEARNING AND MOTIVATION IN
BIOLOGY AND LIFE SCIENCE GRADES 7 THROUGH 12

Improving Understanding and Retention with Writing Notes / Organization

- Significant cogitative effort
- Requires almost two times more effort than from a text
- More thinking effort than playing chess (*they say*)
- Students become more attentive
- Relates to being more active than passive thus decreases issues
- Significant learning takes place
- Provides ownership provided proprietorship/own handwriting makes it personal/Doodles help make a connection between a concept and image in the brain
- Just providing notes can be counterproductive / "skeletal notes/outline" would be better
- Accordion binder

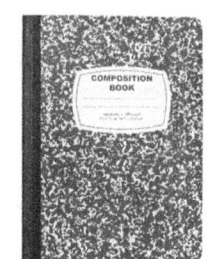

Improving Understanding and Retention with Study Skills

- Always offer study expertise to your students.
- Continue to provide suggestions throughout the year.
- Stress that even gifted students need skills.
- Make suggestions available at all times.
- Mention that studying varies for different people.
- Tell students that procrastination is not beneficial.

Favorite Study Strategies for Students

- Don't be absent from class or be attentive while in class.
- Be prepared before coming to class.
- Correctly do the classwork that is assigned to you.
- Write it all down during lecture.
- Attend study sessions if offered.
- Communicate with your teacher.
- Manage your time. Get sleep. Don't cram. Study daily.
- Know the vocabulary in order to speak the topic.
- Use flash cards for memory reinforcement.
- Only study items that you don't understand.
- Don't try to multitask when studying.
- Try teaching others.
- Practice with textbook, study guide, or lab questions.

White blood cells

Never Let Monkeys Eat Bananas

White blood cells (% of all WBCs)
» Neutrophils (65%)
» Lymphocytes (25%)
» Monocytes (6%)
» Eosinophils (3%)
» Basophils (1%)

- Sample memory techniques—an age-old art (examples):
  - Mnemonics (previously mentioned)
  - Chunking (i.e., discovery of DNA in 19, 53)
  - Memory "palace" (i.e., your home to remember steps of mitosis)

Note: Must practice and the odder the better.

Improving Understanding and Retention with Review Sessions

- Encourage students to attend
- Stress the benefits of the sessions
- Collaborative discussions
- Addressing similar concerns
- Opportunity to narrow down what to study
- Guidelines about what is on the exam
- Before or after school
- Outside school—more advanced classes

Addendum

Other Strategies for Success

Note-Taking Guide

Taking notes in class is one of the most effective ways to understand the material being presented in class. Unless you have a photographic mind, you'll need to learn this important skill.

SCIENCE STRATEGIES TO INCREASE STUDENT LEARNING AND MOTIVATION IN BIOLOGY AND LIFE SCIENCE GRADES 7 THROUGH 12

1. *Come to class prepared.* Always bring enough paper and a writing instrument of your choice to class.
2. *Start a new page for each new class.* Also, put the date on the top of the first page. This way you will know where the notes for each class begin, which will help you keep the material organized. Consider keeping your notes organized in their own binder.
3. *Don't try to write down every word your teacher says.* You will not be able to even if you can write very fast. More importantly, in trying to do so, you will miss the overall point your teacher is trying to make.
4. *Write down the big ideas.* Listen for facts, connections, and main ideas. This may take a while to get used to because you will need to divide your attention between listening to the teacher (or other students) and writing your notes. Don't get frustrated. In time, this will become easier.
5. *Use abbreviations for commonly occurring names and words.* You can develop your own abbreviations, as long as you don't forget what they stand for. For example, in a lecture on Einstein, you might write his name out the first time, and then abbreviate it as "E." throughout the rest of your notes. Long words such as endoplasmic reticulum can become ER. Develop your own system and stick to it once it works.
6. *Leave lots of room on the page.* When writing, leave ample space between ideas. This is like pausing before you begin a new sentence. Your notes will be much easier to read; and you'll have space to add information later on, if needed. Don't try to cram everything onto one piece of paper.
7. *Use diagrams and pictures where necessary.* Sometimes it is helpful to draw pictures that illustrate the connections between ideas, sequences, or events. Don't be afraid to draw pictures that will help you understand the material.
8. *Write down corresponding page numbers from your textbook.* Teachers often use the textbook to refer to ideas you're learning in class. Recording the page number of corresponding ideas and homework assignments can come in handy later on.
9. *Review your notes for accuracy.* It's a good idea to look over your notes sometime after class for accuracy and completeness. Consider doing this just before doing your homework to get yourself back in the mindset of the material.
10. *Obtain notes for missed classes.* Sometimes it's necessary to miss class, but that shouldn't stop you from getting notes for it. Consider forming a partnership with another student at the beginning of class on whom you can rely (and who can rely on you) for notes when a class is missed. Your teacher may also be willing to share his or her notes with you.

Homework Guide

You can improve your performance in classes by ensuring that your homework is understood and completed.

1. *Find an appropriate and consistent environment for study.* This place should be quiet and have adequate lighting and a desk or writing surface. If possible, use this place as a study area exclusively. Avoid distractions, such as TV, stereo, cell phone, internet, video games, etc.
2. *Set up a routine.* Set up a homework routine for the same time daily, if possible. This will let your parents know that you have expectations with regards to homework. It will also help you develop a habit of completing schoolwork at the same time and place daily.
3. *Have your parents help prepare an area for work.* Stock your homework center with the appropriate supplies, such as a dictionary, thesaurus, textbooks, pencils, paper, calculators, etc.
4. *Set little goals for yourself.* Set short-range goals prior to beginning homework. For example, (1) Review the concept of slope, (2) Complete the assigned math homework, and (3) Write a thesis statement for my essay. Take short breaks between assignments for different classes.
5. *Get organized and don't procrastinate.* Use a daily planner and find someone to demonstrate how to use it to organize assignments, extracurricular activities, and other commitments. Keep a three-ring notebook for each subject taken.
6. *Monitor your own progress.* Touch base with your teacher and make sure you have completed all the work. Try keeping your own grades in a notebook. Be aware of what needs to be accomplished each night and check to see that it is done, if necessary. Avoid having your parents or teachers having to hover over you throughout the year.
7. *Confirm that your homework is or was done correctly.* Don't wait until class starts. Go ask your teacher for help well before class so that you and your teacher can discuss your answers appropriately.

Exam-Taking Guide

If you were to prepare for a test the way you might prepare for a big game, you'd probably make a game plan. Okay, so the test probably isn't nearly as much fun as a game, but your goal is the same—develop a plan to win. Being a "bad" test taker is not an excuse to not do well. Often poor test takers don't prepare or prepare correctly when taking an exam.

Before the Exam (see General Study Hints and Guides)

1. *Prepare for the exam by studying for the type of exam given.* If it is a multiple-choice exam, create flash cards that help you memorize the material. If you must write an essay, create outlines that help you see the relationships in the material.

SCIENCE STRATEGIES TO INCREASE STUDENT LEARNING AND MOTIVATION IN BIOLOGY AND LIFE SCIENCE GRADES 7 THROUGH 12

2. *Don't wait until the night of the exam.* Studies show that reviewing the material on a daily or every-other-day basis will help you retain the information better. Waiting to study two to three weeks of material the night before an exam can develop anxiety. Plus, you won't have the opportunity to talk to your teacher about concepts or terms that you're unclear about. Cramming usually does not help with your long-term memory.
3. *Take advantage of the teacher's tutoring or study sessions.* Getting one-on-one instruction can make a world of difference, and going to study sessions may help narrow down the information to be tested. Proper preparation may help reduce anxiety.
4. *Get a good night's rest prior to the test day.* Also, eat a healthy breakfast or lunch on the day of the exam. (Don't overeat!)

When You Begin

1. *Take a deep breath to relax.* Anxiety may reduce your confidence and be an obstacle to doing your best.
2. *Preview the whole test briefly before you begin.* This will help get you warmed up to take the exam and allow you to note the way the exam is organized.
3. *Find out how much time you have to take the test and how much each section is worth.* Allocate your time accordingly. Don't spend the whole test on a section that is worth, say, only 10 points if the exam is worth 120 points.
4. *Read the directions for each section.*

During the Test

1. *Always read the directions before you work on a section.* By reading the directions, it will allow you to make clear decisions on what to do and how to do them. Look for key words such as *and, or, explain*. A lot of times, these words are used to add a second or third part to the question. Failing to read directions can cause you to completely misjudge what the test is asking.
2. *Ask your instructor to explain directions you don't understand.* It is also important to mention that if a question is unclear, ask your instructor to clarify if they can.
3. *Divide and conquer!* Answer the easy questions first to build confidence. This will also allow you to rack up as many points as possible right from the start. However, always be sure to mark the questions you don't answer right away so you can go back to them.
4. *Pace yourself.* Check the time to make sure you're pacing yourself appropriately.
5. *When in doubt, guess.* You at least have a chance that you might guess correctly. *Never* leave a question unanswered even if you have to guess. It might be correct or earn you at least a few points. An unanswered question will be a *zero*.
6. *Don't let others distract you.* Focus only on your own exam. If others are writing and you aren't, don't panic. If others finish before you do, try not to get nervous.

7. *Use any extra time to first make sure you've answered all the questions.* Then, go over the more difficult questions and read them twice. Read essays carefully for accuracy first and grammar second.
8. *Don't change your initial answer unless you have a good reason to do so.* Research indicates that three out of four times, a first choice was probably correct.

Helping Parents

Young people are developing emotionally, intellectually, and physically. Parents can help their child be successful students by encouraging them in the following suggested ways:

1. *Create a quiet space for homework to be completed.* A place free from disruption and fully stocked with supplies is ideal. Do not have your child multitask by studying *and* listening to music, texting, or watching TV.
2. *Communicate with your child.* Ask about their homework and what large projects and tests are upcoming. Also, stay in touch with school. Attend parent-teacher conferences. Ask for periodical updates on your child's progress by contacting the school or educator. Be involved with activities at the school. Show an interest in your child's education.
3. *Play with your child.* Your child should be a best friend of a sorts. Do some things that they like to do, even though you would rather do something else, and put time aside to be with them and/or their school friends. By doing so, you'll develop a bond and get to know your child's lifestyle.
4. *Remember that intrinsic motivation leads to greater creativity.* Allow your child to try a variety of activities (clubs, sports, activities) in order to find their true interests.
5. *Encourage your child to keep trying when faced with a challenge.* Remind your child that teachers and counselors are available to provide extra help both academically and emotionally. Do not belittle topics, such as math or writing, because your child is having difficulty with these topics.
6. *Tell your child to seek help if they need it.* Teachers are happy to meet students' request for tutoring or mentoring. Often, a one-on-one meeting can make a world of difference in learning because of the individual attention. Teachers can set up times for tutoring (i.e., just before an exam) in accordance to the student's schedule for more flexibility.
7. *Encourage your child to read.* If your child wants to improve their scores on standardized tests, reading helps. The best way to improve reading comprehension and vocabulary is to read a lot.
8. *Help your child to set realistic goals and work toward them systematically.* Also, no matter what the goal (making a sports team, improving academic grades, or learning a new skill), remind children that the journey is as important as attaining the goal.
9. *Help with their understanding.* Be there to assist them with their homework or studying. If the subject is unclear, ask your child to explain the topic, communicate with the teacher (phone, e-mail, notes) to help understand the concept, or even use online resources.

10. *Praise your adolescent for their contributions to family, school, and community.* This conveys a belief in their accomplishments and helps to build a positive self-image.
11. *Participate in parent conferences.* It is interesting that parents and guardians stream to parent conferences, meetings, PTOs, etc. while their child is in elementary or middle school; however, once the child is in high school, that participation stops. Is it because parents are getting tired, parents are overwhelmed, parents are making their child more responsible for their actions? Never the excuse. A representative from the family should always attend conferences to primarily stay in communication with the teacher. From elementary to high school (and perhaps after graduation), children still need some form of guidance and support.
12. *Be aware of the bad habits.* Children today have a lot of opportunity to develop habits that are counterproductive if taken to the extreme (i.e., poor diets, lack of exercise, video gaming, TV, late nights). It is tough to manage such habits, but don't give up trying to lower or totally avoid lifestyles that can distract from their studies and plans for the future.

Promoting Greater Student Success by Managing the Biology / Life Science Classroom

Managing in Ways that Promote Greater Success with Information/Policies

- Provide class information with policies (minimal yet thorough).
- Avoid impossible expectations.
- Give yourself wiggle room (i.e., "may..."). Give yourself options.
- Feel free to include consequences.
- Post the policies near a clock.
- Review class expectations midyear.

SCIENCE STRATEGIES TO INCREASE STUDENT LEARNING AND MOTIVATION IN
BIOLOGY AND LIFE SCIENCE GRADES 7 THROUGH 12

Addendum

Biology Objectives, Materials, Requirements, Information, and Expectations

Objectives

- To develop a foundation in biological vocabulary and principles
- To develop an awareness about the uniqueness and diversity of life
- To develop an appreciation of the interrelationships among living organisms
- To develop an understanding of some natural laws and their applications to life
- To develop the concept of commonality of structure and function in living organisms

Axioms of Biology
(Starting Points of Reason)

Cells
Heredity
Evolution
Energy
Regulation

Careers in biology: http://www.aibs.org/careers

Materials and Requirements

Daily Requirements

Composition notebook (only)
Seven-plus (7+) pocket expanding file folder (for handouts, assignments, labs, tutorials, etc.)
Pencil or pen (otherwise, 25 cents toward BioClub, not red or gel)
Assignment(s)

Other Requirements

Chromebook (*charged and ready*)
eText ([paper Version upon special request only] TBD)
Colored pencils (24+ colors, high quality)
Extra paper/pencil/pen
Calculator
Gloves

Please be aware that you may be asked to bring in certain items for labs and/or projects. Be responsible in bringing these items in especially if the task was agreed upon by you.

Class Information and Expectation

* Classroom website: xxxxxxxxxx
* Work e-mail:
* Availability: Around 7:15 a.m. to 7:45 a.m. and 3:10 p.m. to? or e-mail (check class-tutorial schedule outside door, website, Google Drive)

Information

* Be aware of all emergency exits and procedures (fire, tornado, earthquake, lockdowns).
* Communication is highly stressed. Check e-mail and Harmony often. Stop in or visit. And sign up on Remind.
* Failing the semester will *not* be an option. You alone are responsible for your own actions, decisions, and choices.
* Shared Google Drive folders will contain most assignments and tutorials for your convenience, viewing, and retrieval.
* Class study sessions will typically be during or after school. Take advantage of them and any tutorial resources provided.
* Any comments, viewpoints, references, or discussions involving personal religious beliefs are not allowed m class (first amendment).
* Most homework assignments will be for minor points and used for enrichment, checking for understanding, or formative assessment.
* Extra credit is highly limited. See classroom website for an online extra-credit assignment. No extra credit after around eight weeks or a nine-week period.
* The point system will be employed. There will be no weighing or curving of grades. Discretionary points at the end of the nine-week term may be given if a student shows a willingness to try. Get help. Ask questions. Respond to feedback and/or show growth. Exams will count for most of your term points/grade (i.e., midterm and final exam will count 20% of the semester grade). Grades will mainly be based on core-criterion-based standards and primarily by objective performance assessments without bias.

Expectations

Follow all school guidelines.

* Permission and a hall pass are required before leaving the room.
* Include your first and last name and period number on all assignments/exams.
* Refrain from bringing drinks, food, chargers, book bags, coats, and purses to class.

- * Be compliant, prepared, courteous, ethical, attentive, professional, considerate, and pleasant.
- * All work is due when announced; otherwise, it may not be accepted or, on rare occasions, a 10% deduction.
- * Take thorough lecture notes, develop an efficient management system, and employ effective study strategies.
- * Alternative work, limited privileges, detention, time-out, office, parent notification, etc. may all be used as disciplinary actions.
- * Chromebooks are for class purposes. Unless allowed, any electronics seen or used during class time may be confiscated up to 24 hours.
- * Not being in the classroom or in your seat just before the bell or going to the restroom or the drinking fountain during class will be a tardy.
- * Do not be absent from class. If absent, you are required to contact me *outside* class (i.e., via e-mail or before school) and not just prior to or during class; otherwise, it will be assumed that you are ready for class and any assignments when you return. Be aware that a reduction in participation points may result from your absentees. Makeup exams, labs, assignments, and/or activities may be subject to an alternative assignment in the form of an essay or another task, excused or not. The lesson plan outline will help give you an idea of what we are doing in class on a particular day.

Respect and Trust "Number 1 Policy"

By being enrolled in this course, you are agreeing to the said requirements, information, and expectations which may be subject to change.

Biology Club

Open to all biology students willing to truly participate and be active in science and the community. See classroom website, SW Biology Club website, or SW Biology Club Facebook for more information.

Managing in Ways to Avoid Having to Discipline

- Show no favoritism as hard as this may be.
- You have lost if you're yelling. Use soft/firm voice.
- Do not skew from your policies. Maintain a routine.
- Be willing to compromise if necessary.
- Don't sweat the small stuff.
- Use low profile intervention.
- Have a backup plan. Keep them guessing.
- Do not embarrass or single out.

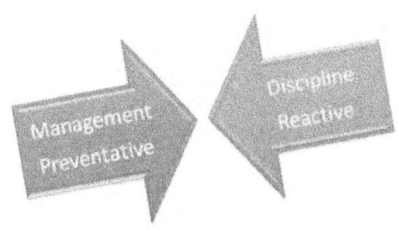

- Avoid sarcasm (toxic humor) and snide/malicious remarks.
- Attention-getters (i.e., hotel bell, wireless doorbell, visual magnets, the "teacher stare," etc.)
- Explain the "cancer analogy" or the "stop signs are for dummies" scenario.
- Anticipate and nip it in the bud.
- Follow up on issues (i.e., after class or the next day).
- Walk among the masses.
- Position yourself accordingly throughout the room.
- Acknowledge compliance and obedience.
- Be sure to document, document, document.

Try to avoid punishment that skews or falsifies standard-based curricula assessment (i.e., zero for no name on an assignment).

Managing in Ways that Promote Greater Success with Trust and Respect

- Do it for yourself and them and they'll do it for one another and you.
- Emphasize how important this policy is to you.
- Be willing to negotiate and compromise.
- "Of all the policies, in my opinion…" = Number 1

Using Demos and Unusual Materials to Increase Understanding of Biology / Life Science Concepts

SEWER LICE (LIFE characteristics / observation / proper background / peer review): Without students noticing, place a few raisins in ginger ale using a clean beaker. Place the beaker on top of an overhead machine and ask students to write down lifelike characteristics in one minute (no talking). After a minute, discuss the characteristics, but then drink/eat the "sewer lice." Now discuss the importance of good observations. One may supplement with a made-up handout on sewer lice to discuss the importance of good research too.

Gregor Mendel (history/genetics): Helps bring the character to life. Custom can be bought on the internet, or wait until Halloween (of just after) to find a "monk" outfit. One may wish to be aware of his biography prior to portraying him (i.e., *Gregor Mendel* by Simon Mawer).

Sugar Snake: Use powdered sugar and sulfuric acid to demonstrate that molecules, such as sucrose, are made up of the elements C, H, and O. Poisonous sulfur dioxide is produced. Do this in well-ventilated area.

Pasteur's Flask (history / spontaneous generation): Bend a glass tube accordingly and place part of it through a hole in a rubber stopper. Place the stopper on a flask containing chicken broth using Vaseline. Do this several months before showing students. As time goes by (years), it will be even more impressive.

Prisoner Escape Trick (scientific method): Have about a meter of string per student. Have them tie their loops around one another's wrists one at a time; however, one string must go over around their partner's string. I suggest having them place a finger between their string and wrist while making the knot for their loop. Give them a problem question, "How can one escape from their partner without removing or breaking the string?" Have them try for a while; however, remind them of the order of the scientific method. For those that ask for a resource, give them a copy of the directions. When done, remind them how hard it was to get out of the trick without researching first and that their many hypotheses were poor because their hypotheses were not established correctly in that background is necessary first before formulating a hypothesis.

Crocheted Models: Provide visuals that are handcrafted.

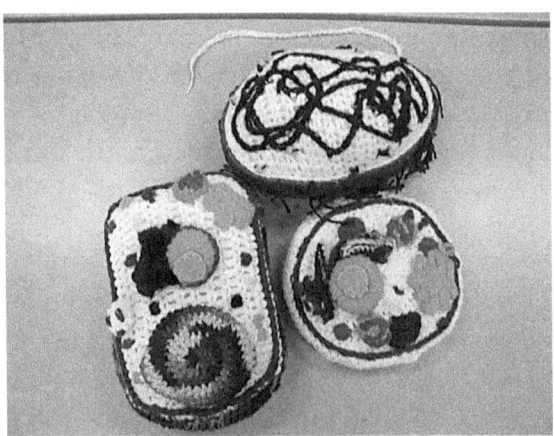

Photosystems 1 and 2 (photosynthesis): Use an old TV dish and crafts to provide visual aid.
DNA Model: Provide a visual aid.
PVC Piping (tracheid and vessel members): Sieve tubes and companion cells in plant tissue.
Punnett Square Socks (intro to genetics): Socks come in pairs like genes.

Cell Membrane Model: Students lay down toe-to-toe in tow rows; thus, representing a phospholipid bilayer.
Phone Tree Evolution: Introduce evolution using words/phrases that change or mutate from one person to another (generations).

SCIENCE STRATEGIES TO INCREASE STUDENT LEARNING AND MOTIVATION IN BIOLOGY AND LIFE SCIENCE GRADES 7 THROUGH 12

Glitter: The STD of Crafts: Using glitter to represent pathogens that can cause STDs, students will be asked to begin shaking hands with two to three volunteer students in order to demonstrate how STDs can be spread (STIs [sexual transmitted infections]).

Pet Store: Several real visual aids for discussing tissues and zoology

Toys (golf clubs [myosin filaments]): Spiky soft ball or pin point impression (cilia), Hula-Hoop (cell cycle), plastic pins and pool noodles (chromosomes), stuffed commercially made cells, toy chicken (DNA from liver)

Props: Toy food, animals, fruits, vegetables, flowers, leaves, etc.

Crafts: Fake fur (mammal characteristic), Halloween web (food web), oven mitts for snail genetics, letters, yellow plastic hat, cupcake mitt (nucleotide)

Specific Topics: Plastic tree (evolution); grape ball (blastula); plastic oven mitt (cells/tissue); practice zippers (DNA); liquid bubble tumbler (ERs/Golgis/Vesicles); Slinky (alpha proteins); Wiffle ball (nuclear membrane); soccer ball (buckyball); button candy (RER); candy canes (DNA); plastic pop links (AAs/proteins); beaded necklace (primary proteins); blocks and plastic pin (enzyme / substrate /active site / reactions); signing the cell cycle (try with hand jive song); Bubble Wrap (cells/tissues); probability box with mousetraps (infantile form of Tay-Sachs demo): rope, chromosome "p" and "q" arms, kinetochore belt, histone pj's, dyads, chromatids (human chromosome); Twist Up cans (DNA/mutations); origami (DNA); pipe cleaners and beads (DNA); commercially made nucleotides (DNA); pet store toy rat to introduce rat dissection and toy chicken with chef's hat to introduce DNA extraction from chicken liver; floating arm trick and falling through the floor sensation (involuntary muscle exercises); flaming Brazil nut (unsaturated fats still have a lot of potential energy); Barrel of Monkeys and chain (hydrogen bonding with water and transpiration); disco mirror ball (X-ray crystallography); different diameters of graduated cylinders with food coloring (vessel members / tracheids in plant tissue); sponge in root shape (root absorption ← turgor pressure); balloon, food dye, peanut with plastic bag, perfume, salt/corn shaker, sugar cube, petri dishes, red poker chips (matrix, gradients, diffusions, molecular, and bulk transport with cells)

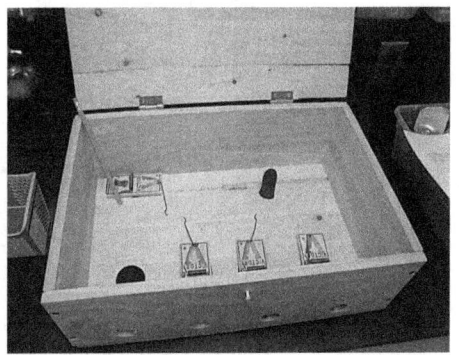

Medical: Donated X-rays, epidural and amniocentesis kits (see eBay), human-to-human interface (see Backyard Brains)

Teacher's Pets (just don't cage them): Use them during discussions too

Cartoons, Stories, and Puns: To display or refer to for particular topics, not a *pun*ishment ☺

Others: Various models, specimens, skeletons, science-related T-shirts, music, oddities.

Philosophy of Failing

Have you ever taken the comment "I failed a student this semester or year" literally?

Never Give Up!

Have you ever heard the phrase "You will never fail unless you give up"?

As long as you keep trying, you won't be giving up on your students, and you'll be the envy of your community.

References

Carey, Benedict. 2014. *How We Learn: The Surprising Truth About When, Where, and Why It Happens.* New York: Random House.

"Does Note-Taking Impede Learning?" Higher Education Pedagogy & Policy. 2010. Accessed December 20, 2016.

Edutopia. https://www.edutopia.org/.

Foer, Joshua. 2011. *Moonwalking with Einstein: The Art and Science of Remembering Everything.* New York: Penguin Press.

Heath, Chip, and Dan Heath. 2007. *Made to Stick: Why Some Ideas Survive and Others Die.* New York: Random House.

Lemov, Doug. 2015. *Teach Like a Champion 2.0: 62 Techniques that Put Students on the Path to College.* San Francisco: Jossey-Bass.

McGuire, Saundra Y., and Stephanie McGuire. 2015. *Teach Students How to Learn: Strategies You Can Incorporate into Any Course to Improve Student Metacognition, Study Skills, and Motivation.* Sterling, Virginia: Stylus Publishing LLC.

Palmer, Parker J. 2007. *The Courage to Teach: Exploring the Inner Landscape of a Teacher's Life.* Tenth anniversary edition. Jossey-Bass.

Scaffolding Definition. http://edglossary.org/scaffolding.

"Seven Ways to Talk to Your Students." Accessed December 23, 2016. http://www.nea.org/tools/52814.htm

Tomlinson, Carol A., and Tonya R. Moon. 2013. *Assessment and Student Success in a Differentiated Classroom.* Alexandria, Virginia: ASCD.

"Welcome to the IRIS Center." Accessed December 16, 2016. http://iris.peabody.vanderbilt.edu/.

Whitaker, Todd. 2004. *What Great Teachers Do Differently: Fourteen Things that Matter Most.* Larchmont, New York: Eye on Education.

Notes

About the Author

ON THE FIRST day of school, have you ever thought of your classrooms as newly opened boxes of crayons? I do. Like pencil-sticks of colored wax, the students each have different names, individual characteristics, and various levels of brightness. I set a goal each year to promote not only creativity but to draw out of my students' reasons about why science is so important. As science educators, we not only need to illustrate the importance of knowing facts and terminology; but, also be able to frame those concepts in such a way that students are motivated to want to study and understand biology. When I began teaching, I never thought that I would have the multitude of experiences I have now. I have taught in schools ranging from city to rural, public to private and large to small; not to mention classes ranging from general science to advanced biology. Through these diverse experiences, I have developed a number of strategies that have enhanced student achievement and science appreciation.

Presenting and submitting articles allows me to contribute to education by sharing concepts, activities, and strategies with my fellow educators. The comradery that ensues while interacting with other teachers during presentations helps bring about opportunities to collaborate and improve student learning. The value and benefit of discussing and deliberating ideas allow not only me, but other teachers to evolve as educators, share instructional methods, and explore different means of engaging all students. Being involved with state-level K-16 organizations such as being elected president of both the Indiana Association of Biology Teachers (IABT) and the Hooser Assoication of Science Teachers, Inc. (HASTI) has allowed me to offer guidance, resources, and support to science teachers. As a representative, I have helped improve our means of communication through social media and orchestrate conferences and workshops. By being involved at so many levels of education, I have helped educators be superior disseminators of knowledge and thus help elevate the quality of science education in Indiana.

Many of the strategies that I use in my classroom are employed to help students be successful, motivated, and improve their understanding and retention of a topic. A few strategies that I use to promote success includes allowing them to take ownership in their work, showcasing their assignments, and varying instruction. Student work is showcased by exhibiting their projects on my wall, in hallway display cases, and on our classroom website. Students are also surrounded by an array of posters, models, preserved specimens, and live animals to add to the ambiance of the classroom environment. I also vary instruction by using technology, manipulatives, hands-on activities, classroom discussions, game reviews, wordplay (puns, analogies, and metaphors), and scaffolding.

To increase their desire to learn, I have implemented strategies such as portraying inviting body language, having an inspiring voice, and presenting with passion when addressing a subject are all keys to instill motivation in the classroom. It is vital to have a growth mindset when addressing how to improve understanding and retention. A few strategies that I have developed beyond the most basic levels of instructional delivery include the use of music, humor, mimicry, mnemonics, familiarities, and memory techniques.

Besides the fact that the core strategies, lessons, and examples have all been conducted by myself, it is stressed that not all the activities need to be performed by the reader. My primary goal of this text is "take away"; to provide ideas to help supplement, expand, reflect on what the experienced or beginning science teacher is now doing in "their" classroom. One of the biggest differences about this text compared to other books is that it is specific to biology and life science educators. This is not to say that any educator can benefit from the concepts mentioned especially with regards other strategies for memory techniques, classroom management and educational philosophies. There are no comparable books that address explicit strategies for biology and life science educators to motivate student to the degree that this text does. Other books may address techniques but at a holistic or all-inclusive scale for "any teacher", which is fine; but not just specific for science educators.

www.ingramcontent.com/pod-product-compliance
Lightning Source LLC
Chambersburg PA
CBHW081718070525
26330CB00036B/707